NUREG-1823

U.S. Plant Experience With Alloy 600 Cracking and Boric Acid Corrosion of Light-Water Reactor Pressure Vessel Materials

Manuscript Completed: March 2005
Date Published: April 2005

Prepared by
B. Grimmel

W. H. Cullen, Jr., NRC Project Manager

Prepared for
Division of Engineering Technology
Office of Nuclear Regulatory Research
U.S. Nuclear Regulatory Commission
Washington, DC 20555-0001

ABSTRACT

This report includes a summary of foreign and domestic Alloy 600 cracking experience, an analysis of the Alloy 600 cracking susceptibility model for vessel head penetration (VHP) nozzles, and information on corrosion of pressure boundary materials in boric acid solutions. This report combines information that the U.S. Nuclear Regulatory Commission (NRC) has collected in response to Davis-Besse Lesson Learned Task Force Recommendations 3.1.1(1) and 3.1.4(1). The survey of Alloy 600 cracking suggests that Alloy 600 and its associated welds (Alloys 182 and 82) are susceptible to crack nucleation and growth in a wide range of applications. The intent of inspection is to identify and remediate the cracking before it can challenge safety systems. VHP nozzle inspection results indicate that the effective degradation years (EDY) model is efficacious and does not suggest a need to revise the model in the short term. However, the EDY calculation is becoming less significant as a result of reactor vessel head replacement. Both the NRC and the industry have implemented boric acid corrosion test programs. Pieces of the Davis-Besse reactor vessel head have been removed and shipped to Pacific Northwest National Laboratory, where they will be examined. The results of the NRC's boric acid corrosion test program at Argonne National Laboratory have shown that the galvanic difference between A533 Grade B Steel, Alloy 600, and 308 stainless steel (used in reactor pressure vessel cladding) is not significant enough to consider galvanic corrosion as a strong contributor to the overall boric acid corrosion process. In addition, the NRC's test program has revealed that the corrosion rate of A533 Grade B steel in humidified salts of H-B-O (a mixture of boric acid species and water) can cause wastage at rates as high as those for saturated aqueous solutions. The industry program, headed by the Electric Power Research Institute, is expected to be completed in 2006.

FOREWORD

Nickel-based alloy cracking of reactor pressure boundary components has been a worldwide concern for about 25 years. Increased inspection frequencies, improved inspection practices, and increased licensee vigilance continue to identify nickel-based alloy cracking in vessel penetrations and various components of the primary coolant loop. Since the U.S. Nuclear Regulatory Commission (NRC) issued Bulletin 2002-02, "Reactor Pressure Vessel Head Degradation and Reactor Coolant Pressure Boundary Integrity," in August 2002, the identified cracks have not been safety-significant. However, leakage resulting from through-wall cracks in these components, leaking joints, valve packings and flanges may also result in boric acid leakage and corrosion. Awareness of the safety-significance of corrosion of low-alloy steel components was raised to a very high level with the discovery of gross vessel head corrosion at the Davis-Besse Nuclear Power Station. To address the issue, the NRC's Office of Nuclear Regulatory Research (RES) sponsored research on the parameters of boric acid corrosion and the corrosion rates of reactor materials.

This report has been assembled from three existing reports. The purpose of this publication is to (a) summarize the current status of plant experience regarding nickel-based alloy cracking and boric acid corrosion, (b) discuss the effective degradation year (EDY) model used in NRC Order EA-03-009 for the prediction of stress-corrosion cracking in Alloy 600 components, and (c) summarize related boric acid corrosion research programs. This report summarizes some important observations, which are detailed in the supporting documents. First, incidents of boric acid corrosion of low-alloy steel reactor components have been frequent, despite a series of guidance documents issued by the NRC. The NRC's guidance culminated with the issuance of Order EA-03-009, "Establishing Interim Inspection Requirements for Reactor Pressure Vessel Heads at Pressurized-Water Reactors," with the intent to stem the findings of boric acid corrosion emerging from the nuclear power plants. The NRC has also developed similar guidance for lower head inspections and inspections of other reactor components. Second, this report concludes that stress-corrosion cracking of nickel-based alloy components, especially those exposed to pressurized-water reactor (PWR) environments, may be expected to continue. However, the inspections required by Order EA-03-009 will ensure that cracking is identified in reactor vessel heads before it compromises structural integrity, and leakage is detected before boric acid corrosion can occur.

As a result of the research funded by RES, together with ongoing research funded by the Electric Power Research Institute, the database of boric acid corrosion effects is more robust than it was a decade ago. In view of the continuing potential for leakage-driven boric acid corrosion, the NRC has supported, and continues to support the American Society of Mechanical Engineers in the development and approval of Code Case N-722, "Additional Inspections for PWR Pressure-Retaining Welds in Class 1 Pressure Boundary Components Fabricated with Alloy 600/82/182 Materials."

Carl J. Paperiello, Director
Office of Nuclear Regulatory Research

CONTENTS

Appendix

Figures

Tables

EXECUTIVE SUMMARY

On March 5, 2002, during a shutdown for inspection and refueling of the Davis-Besse Nuclear Power Station, the plant's licensee, FirstEnergy Nuclear Operating Company, discovered severe boric acid corrosion wastage of the reactor pressure vessel (RPV) head. A triangular cavity approximately 5 inches wide, 7 inches long, and completely through the low-alloy steel RPV head thickness was located downhill of vessel head penetration (VHP) Nozzle #3 approaching Nozzle #11. The wastage consumed between 40 and 60 cubic inches of vessel head. This significantly compromised the integrity of the reactor coolant pressure boundary by reducing it in the vicinity of the boric acid corrosion attack to a section of stainless steel cladding ranging in thickness from 0.199 inch to 0.314 inch. The cladding had an exposed surface area of about 16.5 square inches and had deflected outward under system pressure. Additional boric acid wastage was discovered near VHP Nozzle #2.

This report includes a summary of foreign and domestic Alloy 600 cracking experience, an analysis of the Alloy 600 cracking susceptibility model for VHP nozzles, and a collection of information on corrosion of pressure boundary materials in boric acid solutions. The VHP nozzles that cracked at the Davis-Besse plant were made the nickel-base alloy known as Alloy 600. The survey of Alloy 600 suggests that Alloy 600 and its associated welds (Alloys 182 and Alloy 82) are susceptible to crack nucleation and growth in a wide range of applications, and cracking has not been limited to VHPs. In fact, the survey revealed that Alloy 600 can be expected to crack during the service life of a component. The intent of inspection is to identify and remediate the cracking before it can challenge pressure boundaries and safety systems.

In response to the Davis-Besse event, the U.S. Nuclear Regulatory Commission (NRC) issued Order EA-03-009, which established interim inspection requirements for RPV heads depending upon their effective degradation years (EDYs). The most extensive use of the EDY formula is to predict the occurrence of stress-corrosion cracking (SCC) in Alloy 600 components, including VHP nozzles. Within the limits specified by the Order, the EDY susceptibility model is only applicable to the upper reactor vessel head, but not the lower head because that is at a cooler temperature, with penetrations fabricated differently. VHP nozzle inspection results indicate that the EDY model is effective in designating groups of plants for specific types of inspections, and do not indicate a need for revising the EDY model in the short term. The EDY calculation is becoming less of an issue for high susceptibility plants as a result of reactor vessel head replacement, but is still important guidance for moderate susceptibility plants, which need to examine the trade-off in the costs of increased inspection frequency vs. vessel head replacement. Industry experience suggests that the cost of 100 percent volumetric inspection of heads is greater than the cost of head replacement both financially and in terms of worker dose. Plants that are replacing their heads with Alloy 690 heads will fall into the low-EDY susceptibility category initially and then accumulate EDY as the plant operates with the new head.

Because the Alloy 600 cracking issue has not been fully characterized, there are many foreign and domestic research projects in progress. Test programs for thick-section Alloy 690 and its companion weld metals (Alloys 152/52) are scarce, but several are scheduled to begin in 2005. Additional work on the crack growth rates of Alloys 600, 690, and their associated weld materials is underway in the Westinghouse Science Technology Center. As for the future of Alloy 600 in pressurized-water reactors (PWRs), it is likely that parts, components, and joints fabricated from Alloy 600 and weld filler metals Alloy 82 and 182 will continue to crack during operation. Implementing effective inspection programs will be crucial to identifying and remediating cracking.

Boric acid leakage is a consequence of Alloy 600 cracking. This leakage can lead to boric acid corrosion of the low-alloy steel that is in contact with the aqueous solution or dried solution products. Boric acid corrosion poses a considerable maintenance problem for PWR licensees and, in some cases, has impacted the integrity of low-alloy and carbon steel components. In response to the continued occurrence of boric acid wastage, which has raised concern within the agency and throughout the industry, the NRC has issued several generic letters, bulletins, information notices, and an Order to address this issue. The occurrence of RPV head degradation by boric acid corrosion has changed the emphasis of many inspection plans from leakage-based inspection to crack detection before leakage.

The industry has more than 35 years of experience with this degradation mechanism; however, there is still a large deficit of knowledge with respect to the conditions and mechanisms that produce this phenomenon. The Davis-Besse RPV head wastage that occurred in the years leading up to March 2002 has spurred renewed interest into research on boric acid corrosion and has shown that this process is still not well understood. Both the NRC and industry have implemented boric acid corrosion test programs.

The NRC's Office of Nuclear Regulatory Research (RES) has initiated several programs on boric acid corrosion. Pieces of the Davis-Besse reactor vessel head have been removed and shipped to Pacific Northwest National Laboratory, where they will be examined. At Argonne National Laboratory (ANL), an extensive program of corrosion testing of several boundary materials in concentrated solutions of boric acid has been completed. The results of the NRC's boric acid corrosion test program have shown that the galvanic difference between A533 Grade B steel, Alloy 600, and 308 stainless steel (used in reactor pressure vessel cladding) is not significant enough to consider galvanic corrosion as a strong contributor to the overall boric acid corrosion process. In addition, the NRC's test program has revealed that the corrosion rate of A533 Grade B steel in humidified molten salts of H-B-O (a mixture of boric acid species and water) can cause wastage at rates as high as those for saturated aqueous solutions.

The industry program, headed by the Electric Power Research Institute (EPRI), which is expected to be completed in 2006 was designed with objectives similar to those of the NRC's boric acid program. The results of both the industry and NRC/ANL boric acid test programs will be included in Revision 2 of the EPRI Boric Acid Guidebook (BAGB), which is scheduled for publication in 2007.

ACKNOWLEDGMENTS

The author thanks his colleagues who assisted with the proof-reading and editing of this report. This list includes A. Lee, P. Garrity, E. Sullivan and W. Cullen. Special acknowledgment goes to T. Mintz, S. Crane and W. Cullen, the authors of the documents that provided significant background for this report.

ABBREVIATIONS

ADAMS	Agencywide Documents Access and Management System (NRC)
ANL	Argonne National Laboratory
ANO-1	Arkansas Nuclear One, Unit 1
ANO-2	Arkansas Nuclear One, Unit 2
ASME	American Society of Mechanical Engineers
ASTM	American Society for Testing and Materials
B&W	Babcock & Wilcox
BAC	boric acid corrosion
BAGB	Boric Acid Guidebook (EPRI)
BMI	bottom-mounted instrumentation
BMV	bare metal visual
BWR	boiling-water reactor
CE	Combustion Engineering
CEA	Commissariat à l'Energie Atomique
CEDM	control element drive mechanism
CGR	crack growth rate
CRDM	control rod drive mechanism
DBLLTF	Davis-Besse Lessons Learned Task Force
DNC	Dominion Nuclear Connecticut, Inc.
D/P	differential pressure
ECP	electrochemical potential
ECT	eddy current testing
EDY	effective degradation year(s)
EFPY	effective full-power year
EPRI	Electric Power Research Institute
ET	eddy current testing
GL	generic letter
HPI	high-pressure injection
IAEA	International Atomic Energy Agency
ICI	incore instrumentation
ID	inner diameter
IGSCC	intergranular stress-corrosion cracking
IN	information notice
LER	license event report
MA	mill-annealed
MNSA	mechanical nozzle seal assembly
MRP	Materials Reliability Program (EPRI)
MT	magnetic-particle testing

NDE	nondestructive examination
NEI	Nuclear Energy Institute (formerly NUMARC)
NRC	U.S. Nuclear Regulatory Commission
NSSS	nuclear steam supply system
NUMARC	Nuclear Utility Management and Resources Council (now NEI)
OD	outer diameter
ONS-1	Oconee Nuclear Station, Unit 1
ONS-2	Oconee Nuclear Station, Unit 2
ONS-3	Oconee Nuclear Station, Unit 3
PHB	pressurizer heater bundle
PNNL	Pacific Northwest National Laboratory
PT	penetrant testing (or examination)
PWR	pressurized-water reactor
PWSCC	primary water stress-corrosion cracking
RCP	reactor coolant pump
RCPB	reactor coolant pressure boundary
RCS	reactor coolant system
RES	Office of Nuclear Regulatory Research (NRC)
RIS	regulatory issue summary
RPV	reactor pressure vessel
RTD	resistance temperature detector
RV	reactor vessel
RVH	reactor vessel head
RVLIS	reactor vessel level indication system
SCC	stress-corrosion cracking
SG	steam generator
SSRT	slow strain rate testing
STP-1	South Texas Project, Unit 1
TMI-1	Three Mile Island, Unit 1
TS	technical specification
TT	thermally treated
UHTR	upper head temperature reduction
UT	ultrasonic testing
VHP	vessel head penetration
VT-1	visual testing (or examination)
W	Westinghouse

INTRODUCTION

On March 5, 2002, during a shutdown for inspection and refueling of the Davis-Besse Nuclear Power Station, the plant's licensee, FirstEnergy Nuclear Operating Company, discovered a pineapple-sized cavity in the plant's reactor vessel head. In response to that discovery, the U.S. Nuclear Regulatory Commission (NRC) established the Davis-Besse Lessons Learned Task Force (DBLLTF) to review all related factors and independently evaluate the agency's regulatory processes related to ensuring reactor vessel head integrity. On the basis of its evaluation, the task force developed a report, dated September 30, 2002, which recommended several areas of improvement applicable to the NRC and the nuclear industry.

In November 2002, a senior management review team endorsed all but two of the task force's recommendations. In particular, Recommendation 3.1.1(1) read as follows:

> *The NRC should assemble foreign and domestic information concerning Alloy 600 (and other nickel-based alloys) nozzle cracking and boric acid corrosion from technical studies, previous related generic communications, industry guidance, and operational events. Following an analysis of nickel-based alloy nozzle susceptibility to stress-corrosion cracking (SCC), including other susceptible components, and boric acid corrosion of carbon steel, the NRC should propose a course of action and an implementation schedule to address the results.*

In addition, Recommendation 3.1.4(1) read as follows:

> *The NRC should determine if it is appropriate to continue using the existing SCC models as a predictor of VHP nozzle PWSCC susceptibility given the apparent large uncertainties associated with the models. The NRC should determine whether additional analysis and testing are needed to reduce uncertainties in these models relative to their continued application in regulatory decision making.*

To address the content more reasonably, the NRC subdivided these recommendations into several subtasks. These tasks included (1) development of a survey of Alloy 600 cracking, (2) analysis of the Alloy 600 cracking susceptibility (time-at-temperature) model for vessel head penetration (VHP) nozzles, and (3) collection of information on corrosion of pressure boundary materials in boric acid solutions.

This report achieves the NRC's strategic goal of "openness" by presenting the information compiled from these tasks.

2.0 ALLOY 600 CRACKING

2.1 Introduction

The impetus for collecting information on Alloy 600 cracking was the March 2002 discovery of the vessel head penetration (VHP) flaws, leaks, and pressure boundary corrosion at the Davis-Besse plant (Ref. 1). Since this discovery, U.S. and international reactor operators continue to discover (primarily through the observation of leaking products) cracks in primary boundary piping and penetrations fabricated from wrought or welded nickel-based alloys. Events of recent note are the cracks in a medium-diameter pressurizer nozzle at Tsuruga 2 and the re-cracking of the repaired instrument nozzle in the head that was replaced at Oconee 1 (NRC Event Notification 40192, described further in Appendix A to this report). In the United States, significant findings in components other than VHPs have been observed at V.C. Summer (hot leg "A" cracks, predominately axial but with circumferential components); South Texas Project 1 (bottom-mounted instrumentation nozzle cracks); and North Anna 2 (well-developed, outside diameter, circumferential cracks in a VHP penetration not exhibiting leakage products). By reviewing the history of Alloy 600 cracking, one can conclude that Alloy 600 will continue to crack during the service life of the component. The intent of inspection is identifying and remediating cracking before it can challenge plant safety systems.

2.2 Historical Background

Nearly all reviews of primary water stress-corrosion cracking (PWSCC) in Alloy 600 begin with the mention that this phenomenon was first observed (Ref. 2) in 1959, by Henri Coriou [in the laboratory now known as the Commissariat à l'Energie Atomique (CEA)], in pure, deoxygenated (~ 1MΩ, ≤3 part per billion O_2) water at 350 °C (662 °F). Tests were also conducted in water containing 1 ppt (part per thousand) chloride (Cl) as sodium chloride. Alloy 600 exhibited cracking at 0.5-percent permanent strain after 9 months in pure water and 6 months in chloride-contaminated water. By the early 1970s, the basic dependencies of Alloy 600 PWSCC on carbide precipitation, temperature, and levels of minor contamination were well known. Most of the early work was propelled by cracking in steam generator tubing. Of particular importance is the work by Copson and Dean (Ref. 3), who studied the effects of contaminants, especially lead, on initiation times and rates of crack extension.

Initially, the work of Coriou was challenged by other experimentalists who had no success in cracking Alloy 600. These detractors suggested that inadvertent and immeasurably low levels of contamination may have been responsible for the cracking. Repeated testing and subsequent documents authored by Coriou (Ref. 4) asserted that the CEA autoclave environments were as pure and carefully maintained as possible. Coriou attributed his success in obtaining cracking results to the relatively high levels of strain in the test coupons, as well as a willingness (and ability) to allow tests to run for many consecutive months. Ultimately, many others were able to obtain similar indications of cracking, and the results of Coriou, together with other tests from the 1960s, led to recognition of the effects of strain, temperature, microstructure, yield strength, and certain environmental parameters, including oxygen level, and contaminants such as lead and chlorine. Not every one of the conclusions resulting from this work remained completely intact over the years, but the CEA group developed a remarkably clear picture of the important parameters. By the early 1980s, cracking of steam generator tubing, first on the secondary side, and later from the primary side, provided notice that Alloy 600 would crack in field applications, given a high enough stress and a substantial incubation time.

The tests of Coriou and his contemporaries are best characterized as crack initiation tests. For the most part, the test samples were small flat beams or rings, strained by clamping mechanisms or tie rods, and contained in autoclaves without any additional instrumentation to monitor strain relaxation or crack extension, nucleation, or degradation. By the early 1990s, testing with fracture mechanics specimens (usually of compact design), together with very accurate and stable crack extension measurement systems, had been established in many laboratories worldwide, and well-characterized crack growth rate data became available (Ref. 5).

Cracking in primary side water of thicker sections of Alloy 600 was heralded domestically by the discovery of leaks in pressurizer instrument nozzles (in 1986 at San Onofre Nuclear Generating Station, Unit 3) and heater sleeves (in 1987 at Arkansas Nuclear One). In 1989, 20 leaking heater sleeves were found at Calvert Cliffs, Unit 2 (8 additional sleeves had non-leaking, axial crack indications). A 1990 literature survey documented the history of this and similar cracking discovered in both domestic and French plants (Ref. 6). At this stage of understanding, incidents of field cracking were strongly correlated with temperature; pressurizer penetrations are generally exposed to a higher temperature than VHP penetrations and other components. Figure 1 is a timeline of the significant findings in reactor plants, along with the conference and significant reporting milestones.

The 1990 report yielded the correct conclusion that susceptibility of Alloy 600 was related to its thermomechanical processing. Specifically, a low-temperature final anneal, producing intragranular carbides and a yield strength at the high end of the specification, coupled with a fabrication process (such as cold work or welding) that imposed a high residual stress, led to a high susceptibility to crack initiation and growth.

A 1991 workshop (Ref. 7) focused on the specifics of plant experience and specific incidents. In addition to these sessions on plant experience, other sessions covered inspections, repair strategies, testing and analysis, modeling, and outage planning. The workshop included one of the first presentations describing the stress distribution in the vicinity of a vessel penetration with a J-groove weld attachment on the coolant side. The importance of temperature and material microstructure were confirmed and described further. Other firsts included the identification of dissolved hydrogen level as an important factor; the influences of lithium-to-boron ratio (Li/B) and pH were cited as uncertain.

A larger workshop in 1992 reflected the growing worldwide concern. More than 50 presentations (Ref. 8) provided a much greater level of detail. As a result of the discovery of VHP cracking at Bugey 3 in 1991, the focus of the 1992 conference shifted from pressurizer penetrations to reactor vessel penetrations. The French reported that about 3 percent of the inspected nozzles had flaw indications. Stress analysis and characterization of materials, including weld metals, was much more extensive than in the 1991 workshop.

4

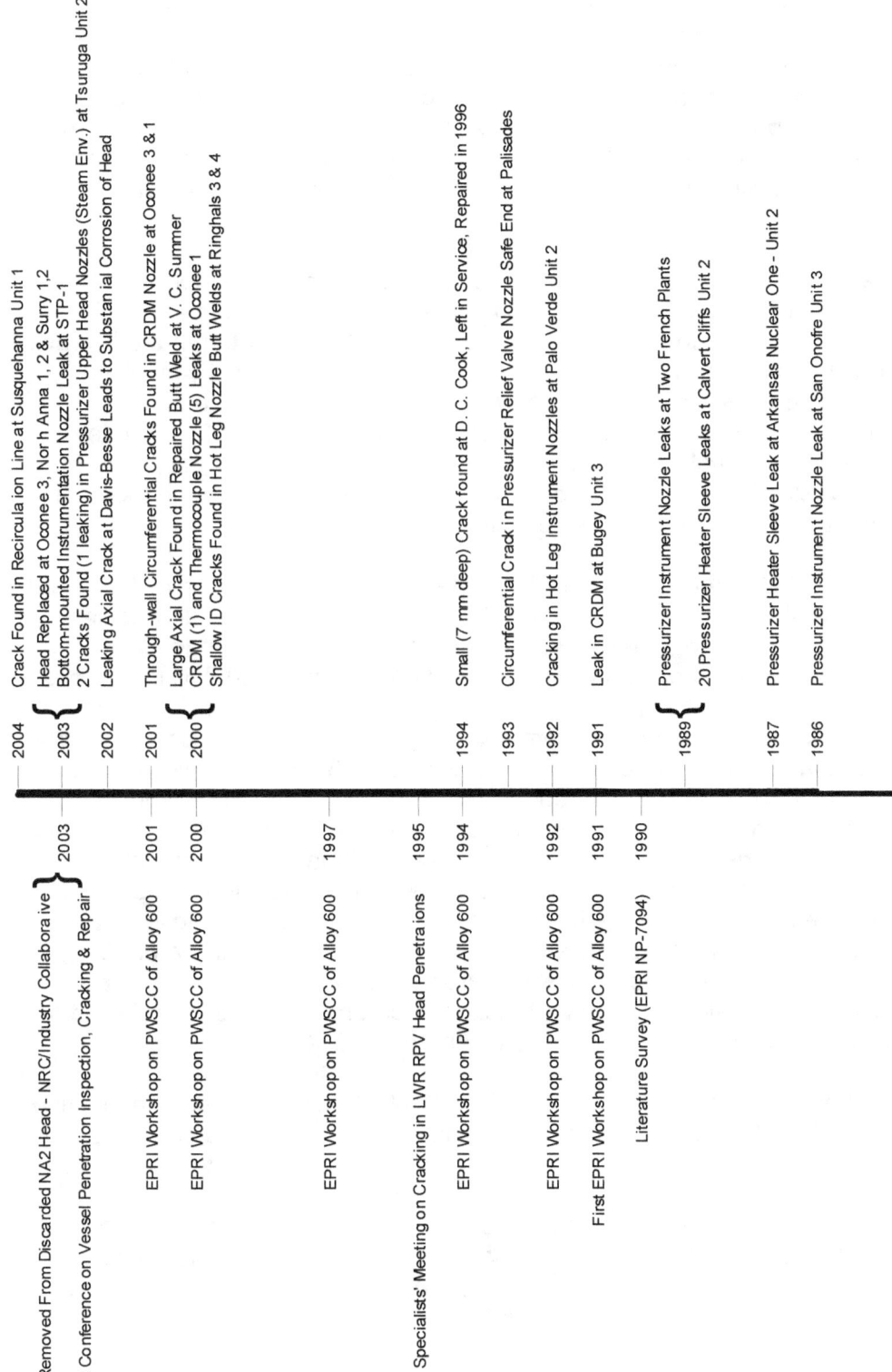

Figure 1. Timeline of findings of significant Alloy 600 cracking in NSSS plants, and important documentation, workshops, and conferences held to discuss this issue.

5

One presentation at the 1992 workshop described the destructive examination and the findings of 11 axial cracks, including one leaking crack, on the inner diameter (ID) of one nozzle at Bugey 3. The leaking crack in Nozzle #54 (one of the outermost locations) was about 50 mm long on the ID and about 2 mm long on the outer diameter (OD). At that time, it was unequivocally stated that no circumferential cracks had been observed. A second through-wall crack was found in the same VHP, but on the uphill side and below the weld. Axial cracks were found in one additional nozzle at Bugey 3; the remaining 63 nozzles were determined to be not cracked. At the sister plant, Bugey 4, cracking with an axial orientation but no leaking was found in eight nozzles.

Examinations of the cracked Bugey-3 VHP housing continued for a long time after (Ref. 8). There was one, small (3 mm long x 2.25 mm deep) indication on the outside of the nozzle above the weld at an angle of 30° off the vertical axis. It was initially postulated that this crack was a branch off the leaking axial crack. However, subsequent presentations by involved French engineers (Ref. 9) suggested that this may have been a crack that nucleated independently of the main axial through-wall crack. There was also a 110° by 3.5 mm deep circumferentially oriented flaw in the weld, on the uphill side, that emanated from the "triple point" (the point at which the low-alloy base metal, J-groove weld, and VHP tube all come together), and propagated into the buttering and the J-groove weld.

The 1994 workshop on PWSCC of Alloy 600 (Ref. 10) expanded on the same topics that were incorporated in the 1992 workshop. In the 2-year interval between the 1992 and 1994 workshops, VHP inspections in several nations (i.e., vessel heads fabricated by several vendors) revealed mostly axial cracks. Some plants with up to 20 years of operation did not exhibit flaw indications in the VHPs, but one plant indicated that an Alloy 600 pressurizer steam-space instrument nozzle, welded with Alloy 182 (as compared with the more common Alloy 152) suffered cracking in the weld. The proceedings reported that vessel head replacement began in 1993–1994. Six vessel heads were replaced on French plants during those 2 years, and 29 vessel heads were ordered. Plans were also announced for replacement of "two vessel heads in Spain, one vessel head in Sweden, and three vessel heads in Japan." It was also reported that computational stress analyses had shown that (1) the development of welding-induced residual stress was dependent on the fabricator, (2) stresses were not always higher for the downhill nozzles (those located toward the periphery of the head), and (3) butt weld stresses varied substantially around the circumference and depended on the specific welding procedure. In addition, it was reported that the PWSCC susceptibility of weld metal was a function of chromium content, and welding alloys with greater than 30 percent chromium were resistant to PWSCC.

The head penetration inspection experience at the D.C. Cook 2 plant was described at a meeting following the conclusion of the 1994 workshop. Specifically, in 1994, three axial flaws were detected in an VHP nozzle (#75). Those flaws were axial in orientation, with lengths of 45, 16, and 9 mm and a maximum depth of 6.8 mm [compared with a wall thickness of 0.625 in. (16 mm)]. All flaws were below the weld, although the largest flaw was very close to the bottom of the weld. A fracture mechanics evaluation indicated that growth of these flaws would not exceed 75 percent of the wall thickness during the next operating cycle. Consequently, they were left in place and repaired two cycles later in 1996.

6

A 1995 conference (Ref. 11), held under the auspices of the International Atomic Energy Agency (IAEA), was one of the first opportunities for the NRC to publicly present the agency's position regarding the safety-significance of the VHP cracking issue. J. Strosnider took the opportunity to point out that leakage and ensuing deposits of boric acid on the head could lead to high corrosion rates. He also suggested that the unique stress distribution developed by the attachment weld could lead to the possibility of circumferential cracking. This conference was also the forum for the IAEA to describe the development of an "International Database on Aging Management." Participants from Spain, France, Korea, Japan, and the United States presented summaries of the vessel head cracking situation or the vessel head replacement programs in their respective countries. The results of 3-D modeling of the residual stress distribution for a J-weld attachment were the subject of two presentations.

A 1997 workshop (Ref. 12) was held in the shadow of the NRC's imminent issuance of Generic Letter (GL) 97-01, which required licensees to describe their plans for vessel head inspection for VHP cracking. There had been no significant change in the incidence of VHP cracking in other parts of the world, and one occurrence in the United States (D.C. Cook, which was repaired in 1996), between the 1995 and 1997 conferences. The French plants had initiated a replacement program by this time, and cracking of pressurizer penetrations in domestic plants continued to be an economic, but not a safety, problem. Nonetheless, the NRC was about to require (in GL 97-01) plants to develop a long-term plan for inspection and monitoring for PWSCC of vessel head penetrations.

Three years elapsed before the next industry-sponsored workshop in 2000 (Ref. 13). The opening presentation by W. Bamford pointed out that reactor pressure vessel (RPV) head nozzle cracking or flaw indications had been found in 11 countries. Cracking had been found in about 6.5 percent of the nozzles inspected in French plants. Eight inspections of bottom-mounted instrumentation (BMI) tubes in four countries had not revealed any evidence of cracking at the time of the conference. Cracking in several repaired pressurizer steam-space nozzles, and laboratory testing of Alloy 600 and several variations of Alloy 81 (viz. Alloy 182, EN 82H), showed that crack propagation rates in the weld metal were higher than in base metal.[1] Some research on crack sizing and mitigation of crack nucleation (by water-peening, shot-peening, nickel-plating, and others) was presented at this workshop.

Cracking in foreign plants continued in 2000. During a 2000 inservice inspection at Ringhals Unit 3 in Sweden, two axial indications were detected in the Alloy 182 weld material between the RPV nozzle and the safe end of the reactor coolant hot leg piping. The deepest crack was reported to be 9 mm into the weld from the inner surface. There were no documented repairs in the areas where the cracks were found (Ref. 15). Additionally, during an inservice inspection at Ringhals Unit 4 in 2000, four axial cracks were found in one outlet nozzle. All of the cracks were in the Alloy 182 weld material between the RPV nozzle and the safe end of the reactor coolant hot leg piping. The depths of the cracks were between 6 mm and 22 mm. All of the cracks were found in repaired areas (Ref. 15).

---

[1] By mid-2003, a sufficient tabulation of crack growth rates for both Alloy 600 and Alloy 182 allowed an evaluation showing that SCC growth rates in Alloy 182 weld metal were a factor 2.7 faster than rates in Alloy 600 well-controlled laboratory tests. A carefully constructed non-linear regression fit to this data, and elimination of the (artificial) "threshold" of 9 MPa· m that was used for the Alloy 600 curve, produces a curve that is virtually indistinguishable (for K > 20 MPa· m) from the 5X "Scott" curve for Alloy 182 that was proposed in MRP-21 (Ref. 14).

The 2001 workshop (Ref. 16) was held following the discovery of several instances of leaks in domestic reactors. Leakage from VHPs had been found at Oconee 1 (Fall 2000 outage), followed by Oconee 3 (early 2001 maintenance outage), and Oconee 2 (Spring 2001 outage). Intermixed with the Oconee findings, a leaking crack was also found at Arkansas Nuclear One, Unit 1 (ANO-1), in February 2001. The "A" hot leg crack had been discovered at V.C. Summer in October 2000. Detailed presentations by all of the affected licensees described their inspection findings and the repair procedures in which they engaged. The sense of urgency created by these findings provoked discussions on whether leakage would always be present, and whether cracks would become a safety concern before they would be discovered by inspection. The importance of maintaining a clean head was particularly stressed in a presentation on the Oconee findings. J. Strosnider gave the regulatory perspective on VHP nozzle cracking, stressing that this issue was attracting senior management attention at the NRC, and that inspection approaches based on leakage detection alone may not constitute compliance with the regulations.

After a 2-year hiatus from conferences and workshops, the NRC, with the assistance of Argonne National Laboratory (ANL), held the "Vessel Penetration Inspection, Crack Growth, and Repair Conference" on September 29 – October 2, 2003.[2] Many of the papers focused on recent developments in the areas of inspection, understanding of crack growth mechanisms, and mitigation of crack growth. The timeliness of the presentations was exemplified by three contributions by the staff at South Texas Project, describing elements of the discovery and repair of their BMI nozzles, and a presentation describing the findings about the pressurizer nozzle leaks discovered 3 weeks earlier at the Tsuruga plant. The presentations in the lead-off technical session on nondestructive inspection technology noted that PWSCC of nickel-base alloys is a complex problem, and few of the key factors are fully understood.

Cracking continued into 2004 at Ohi Power Station Unit 3 in Japan. During an inspection at Ohi Unit 3 in 2004, a deposit was found from the piping nozzle stub for installment of a VHP. The deposit was determined to be boric acid from primary coolant leakage. The piping nozzle stub is welded to the reactor vessel and piping (Ref. 17).

The NRC conference included presentations that described several recent Alloy 600 cracking events. Appendix A to this report contains an up-to-date summary of findings gleaned from license event reports (LERs) submitted by U.S. licensees. The LERs have been included to give a more complete picture, and to expedite traceability. One theme that emerges from this compilation is that, in pressurized-water reactor (PWR) plants, cracking has progressed from pressurizer penetrations (beginning in the mid-1980s), to VHP nozzles (in the 1990s), hot legs (late 1990s), to BMIs (2003). This progression follows a pattern of successive occurrence components operating at progressively lower temperatures. In boiling-water reactors (BWRs), findings of cracking and leakage are just beginning to emerge. Most of the cracks or leaks found to date have been in recirculation lines (in which thermal fatigue may be a component of the driving force), and core shroud supports. It is too early to tell whether a pattern exists that suggests locations of future cracking in BWRs.

---

[2] Proceedings from this conference are in publication and should be available in mid 2005.

2.3 U.S. Plant Experience and Regulatory Actions

Based on cracking in primary side water of thicker sections of Alloy 600 in pressurizer instrument nozzles and heater sleeves in U.S. plants, and VHP nozzles at Bugey, the NRC staff implemented an action plan in 1991 to address PWSCC of Alloy 600 VHPs at all U.S. PWRs. The action plan included reviewing safety assessments from the PWR Owners Groups, developing of VHP mockups by the Electric Power Research Institute (EPRI), qualifying of inspectors on the VHP mockups by EPRI, reviewing of proposed generic acceptance criteria from the Nuclear Utility Management and Resources Council (NUMARC) [now the Nuclear Energy Institute (NEI)], and performing VHP inspections. Upon a review of the industry's safety assessment and an examination of the foreign inspection findings, the NRC staff concluded in a safety evaluation dated November 19, 1993 (Ref. 18), that VHP nozzle cracking was not an immediate safety concern. The rationale for this conclusion was that if PWSCC occurred at VHP nozzles (1) the cracks would be predominately axial in orientation, (2) the cracks would result in detectable leakage before catastrophic failure, and (3) the leakage would be detected during visual examinations performed as part of surveillance walkdown inspections before significant damage to the reactor vessel head would occur.

The industry safety assessments reached several conclusions. One conclusion was that an analysis of circumferential cracking (Ref. 19) revealed that a through-wall circumferential flaw would take more than 40 years to grow circumferentially along the elliptical weld zone toward the uphill location. Therefore, the analysis concluded that there was not a possibility for an external circumferential flaw indication to grow circumferentially to an extent to become a safety concern. Another industry report (Ref. 20) concluded that outside diameter surface cracking has no safety-significance because it would take longer than 90 years for a crack to grow through-wall (after initiation) and propagate circumferentially to a length of 330••

The first U.S. inspection of VHP nozzles occurred in 1994 at Point Beach Nuclear Generating Station. No indications of flaws were detected in any of its 49 VHPs (Ref. 21). An eddy current inspection at Oconee Nuclear Station in 1994 detected 20 indications in a single penetration. In 1994, an examination at D.C. Cook 2 detected 3 axial flaws in one VHP. The flaws were axial in orientation, with lengths of 45, 16, and 9 mm, and a maximum depth of 6.8 mm [compared with a wall thickness of 0.625 in. (16 mm)]. All flaws were below the weld, although the largest of the flaws was very close to the bottom of the weld. Table 1 is a compilation of VHP cracking incidents in the United States since 2000.

On April 1, 1997, the NRC issued GL 97-01 to request PWR licensees to submit descriptions of their programs for inspection of control rod drive mechanism (CRDM) and other VHP nozzles. As described in a generic response to NRC requests for additional information (Ref. 22), the industry used a histogram grouping of plants, in combination with completed inspections and planned inspections as its approach for managing the issue. The plant grouping used probabilistic crack initiation and growth models to estimate the amount of time remaining (in effective full-power years, EFPYs) until the plant reached a limiting condition for a reference plant.

In the Fall of 2000, boron deposits were identified on the RPV head at one VHP nozzle and five thermocouple nozzles at Oconee Nuclear Station, Unit 1 (ONS-1). An analysis of the VHP nozzle identified an axial-radial PWSCC crack that initiated in the J-groove weld and propagated part way into the outer diameter surface of the nozzle. The discovery of nine nozzles with leaks at the Oconee Nuclear Station, Unit 3 (ONS-3), maintenance outage in February 2001 was significant. Two of the nozzles had a through-wall circumferential crack extending 165° around the nozzles, though the cracks were not through-wall for their entire circumferential extent. These cracks initiated from the nozzle's OD surface. This discovery contradicted the industry analysis mentioned above that stated that a circumferential flaw could not grow to the extent to become a safety issue.

Table 1. VHP cracking incidents in the United States since 2000.

Plant	Nature of Cracking, Head Replacement Plans
Oconee 1	Leaks or cracks detected in 3 VHPs + 5 instrument nozzles in 2000 (Head replaced in 2003)
Oconee 2	19 VHPs repaired, 1 circ. crack above the J-weld (Head replaced in March 2004)
Oconee 3	14 VHPs repaired, 4 with circ. cracks above J-weld (Head replaced in 2003)
ANO-1	8 VHPs repaired (Head replaced in 2005)
Surry 1	6 VHPs repaired (Head replaced in 2003)
North Anna 2	14 VHPs repaired, 6 with circ cracks (Head replaced in 2002)
Davis-Besse	5 VHPs with cracks, 2 nozzles with wastage (Head replaced in 2003)
Three Mile Is. 1	6 VHPs and 8 instrument nozzles plugged (Head replaced in 2003)
Crystal River 3	1 VHP with circ. crack (Head replaced in 2003)
Ginna	Head replaced in 2003
Millstone 2	3 control element drive mechanisms (CEDMs) repaired, (Head replaced in 2005)
South Texas 1	2 bottom head instrumentation nozzles repaired
Heads to be replaced at Kewaunee (2005), Robinson 2 (2005), Pt. Beach 1 (2005), Farley 1 (2004), Pt. Beach 2 (2005), Farley 2 (2005), Prairie Island 2 (2005), St. Lucie 1 (2005), St. Lucie 2 (2006), and Turkey Pt. 4 (2005), Heads replaced at Turkey Pt. 4 (2005), North Anna 1 (2003), and Surry 2 (2003)	

In March 2001, ANO-1 detected boron deposits on a VHP nozzle. Examination of the nozzle identified an axial part-through-wall crack that initiated on the nozzle's OD surface below the J-groove weld and propagated to a distance 33 mm (1.3 in) above the J-groove weld. Shortly after this time, in April 2001, Oconee Nuclear Station, Unit 2 (ONS-2), identified boron deposits on four VHP nozzles. Leakage from these nozzles was identified as originating from the outer diameter surface cracks that propagated along the weld-to-nozzle interface from below the J-groove weld to above the weld.

In April 2001, the NRC issued Information Notice (IN) 2001-005, "Through-Wall Circumferential Cracking of Reactor Pressure Vessel Head Control Rod Drive Mechanism Penetration Nozzles at Oconee Nuclear Station, Unit 3" (Ref. 23). As described in this information notice, initial nondestructive examination (NDE) of the vessel head penetrations identified 47 recordable crack indications, which were characterized as either axial, or below-the-weld circumferential indications. Subsequent PT revealed that two of the nine nozzles had "significant circumferential cracks in the nozzle above the weld" (Ref. 23). Both cracks had initiated from the outside diameter of the nozzles; one was through-wall, and the second had "pin-hole through-wall indications." This IN underscored the importance of thorough visual examinations of the reactor head, or volumetric examinations of the VHP nozzles, and appropriate characterization of any flaw indications that might be found.

Subsequently, the NRC issued Bulletin 2001-01, "Circumferential Cracking of Reactor Pressure Vessel Head Penetration Nozzles," (Ref. 24) indicating that the staff had reassessed the earlier conclusions from GL 97-01 — that cracking of vessel head penetration nozzles "is not an immediate safety concern." The bulletin specifically cited the finding of circumferential cracks at Oconee, comparing that with the GL 97-01 assumption that all cracks would be axial. The bulletin stressed that prior deposits of boric acid on vessel heads could mask, or obscure, evidence of small, ongoing leakage. The bulletin pointed out that insulation on the RPV head, or "other impediments" may restrict an effective examination, and reinforced the importance of conducting effective examinations for leaks, head degradation, or cracks. The bulletin laid out the specific paragraphs in the General Design Criteria that specify the criteria for vessel inspection and analysis, and the acceptance standards for any identified degradation. The bulletin also requested that within 60 days of its issuance, all licensees provide descriptions of the vessel head penetrations, head insulation packages, and other aspects of vessel head design pertaining to inspectability. Finally, the bulletin requested additional information from plants that had experienced leakage. When the 60-day responses were reviewed, the staff determined that a request for additional information was necessary to supplement the information originally provided by a few plants.

While conducting inspections in accordance with Bulletin 2001-01, severe boric acid corrosion wastage of the reactor pressure vessel head was discovered at Davis-Besse Nuclear Power Station in March 2002. Figure 2 is a metallograph of the flaw from Nozzle #3 — the location where significant head degradation occurred — from the Davis-Besse Plant. The characteristics of this discovery are detailed in Section 3.5. In response to this event, the NRC issued Bulletin 2002-01, which is described in Section 3.3. Subsequently, the NRC also issued Bulletin 2002-02 (Ref. 25) mainly to advise PWR licensees that visual examinations as a primary inspection method for the RPV head and VHP nozzles may need to be supplemented with additional measures (e.g., volumetric and surface examinations) to demonstrate compliance with applicable regulations.

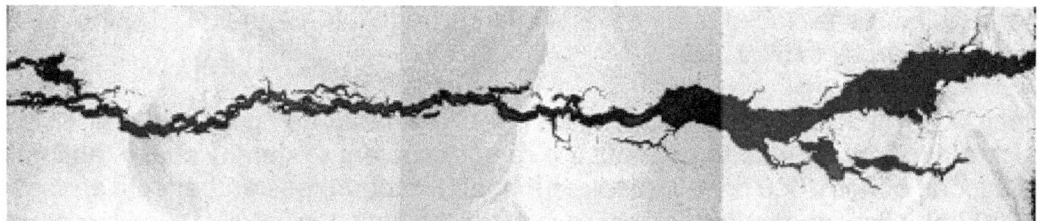

Figure 2. A metallograph showing the profile of a crack in the attachment weld of Nozzle #3 in the (now discarded) Davis-Besse reactor head (Ref. 1).

NRC Order EA-03-009, issued in February 2003 (Ref. 26) and revised in February 2004 (Ref. 27), strengthened PWR inspection plans with respect to the RPV head and VHP nozzles. The continuing occurrences of boric acid corrosion wastage, epitomized by the Davis-Besse incident, have once more brought to the foreground the insufficient inspection criteria for the RPV head and associated penetration nozzles. The inspection requirements in place before Order EA-03-009 required only inspection of the head insulation and surrounding areas for leakage. This type of inspection is not effective for detecting head degradation and circumferential cracking. Therefore, licensees are required to calculate their plants' effective degradation years (EDYs), as described in Section 2.4. The plant susceptibility to PWSCC of the nickel-based

11

vessel head penetrations, based on the EDY, determines which inspection plan dictated by Order EA-03-009 is required to be implemented.

Per the original order and its 2004 revision, the plants were divided into four susceptibility categories: high, moderate, low, and replaced. High-susceptibility plants are those with an EDY greater than 12 years, or those that have experienced PWSCC in nozzle penetrations or the J-groove weld. Moderate-susceptibility plants are those that have an EDY between 8 and 12 years and have not experienced PWSCC in nozzle penetrations or the J-groove weld. Plants classified in the low-susceptibility category have an EDY less than 8 years and have not experienced PWSCC in the base metal nozzle penetrations or the J-groove attachment weld. Lastly, plants in the replaced category are those that have replaced the RPV head and (therefore, by definition) have an EDY less than 8 years and have not experienced PWSCC in nozzle penetrations or the J-groove weld (Ref. 27). Section 2.4 provides a more detailed description of the EDY formula.

Each licensee is required to conduct (1) a bare metal visual (BMV) inspection of 100 percent of the RPV head and nozzle penetrations, and (2) nonvisual NDE of (a) the volume the RPV head penetration 2 inches above the highest point of the root of the J-groove weld to 2 inches below the lowest point of the toe of the J-groove weld by volumetric UT, or (b) the entire wetted surface of the area of the RPV head penetration 2 inches above the highest point of the root of the J-groove weld to 2 inches below the lowest point of the toe of the J-groove weld by eddy current testing (ECT, or ET) or penetrant testing (PT) (Ref. 26). However, the schedule of these inspections varies, depending on the licensee's susceptibility rating.

For plants in the high-susceptibility category, BMV of the head and NDE of the penetrations are required every outage. For plants in the moderate-susceptibility category, BMV of the head and NDE of the penetrations may be alternated every outage. For plants in the low-susceptibility category, BMV of the head must be completed every third outage, and NDE of the penetrations every forth outage. Lastly, for plants in the replaced category, no inspection is required during the outage in which it is replaced and the inspection schedule thereafter will be the same as that for low-susceptibility plants (Ref. 26).

With the focus on PWSCC of Alloy 600 on the upper RPV head and possible boric acid corrosion of ferritic components throughout the reactor coolant system, visual examinations of other applications of Alloy 600 have increased in their thoroughness and effectiveness. In the Spring of 2003, the licensee for the South Texas Project, Unit 1 (STP-1), identified boron deposits on the lower RPV head near two BMI nozzles (Ref. 28).

Characterization of all of the BMI nozzles at STP-1 identified PWSCC in these two nozzles, and no PWSCC in any other nozzle (Ref. 29). Subsequently, The NRC issued Bulletin 2003-02, "Leakage from Reactor Pressure Vessel Lower Head Penetrations and Reactor Coolant Pressure Boundary Integrity" (Ref. 30), to obtain information on licensee inspection activities and inspection plans for the RPV lower head.

The 2004 revision to Order EA-03-009 specifies that if the surface of the RPV head located downslope of the outermost RPV head penetration is obscured by support structures, the BMV inspection must include no less than 95 percent of the vessel head, and if any corrosion products or evidence of boron is detected, then the support structure must be removed and the obscured region must be inspected. In addition, the revision allows for a relaxation in the inspection of inconsequential portions of the RPV head penetrations. If the region greater than 1 inch below the lowest point of the toe of the J-groove weld sees operating stresses less than 20 ksi in tension, then only the volume from 2 inches above the highest point of the root of the J-groove weld to 1 inch below the lowest point of the toe of the J-groove weld is required to be inspected. Furthermore, the clarification was made that a combination of UT, EC, and PT may be performed to screen equivalent volumes, surfaces, and leak-paths of the RPV head penetration 2 inches above the highest point of the root of the J-groove weld to 2 inches below the lowest point of the toe of the J-groove weld (Ref. 27).

The GLs and INs issued in the 1980s and early 1990s stressed the deficiency of UT in detecting flaws and, therefore, promoted a leakage-based inspection plan that relied heavily on a combination of NDE, but primarily visual inspection. NDE and UT analysis methods have greatly progressed since that time and investigations into improved NDE methods are ongoing. While visual inspection, particularly bare metal visual inspection, is still an integral part of the inspection plan, it is being supplemented by other NDE technologies (mainly much improved UT and ET methods) in an attempt to detect flaws before they result in leakage. This trend away from leakage-based inspection is exemplified in EA-03-009.

Cracking continued in 2004 at Catawba Unit 2. On September 16, 2004, the steam generator bowl drain for the 2A, 2C, and 2D steam generators were visually inspected. Leakage was found on the 2C and 2D SG bowls. The leakage occurred sometime after the previous refueling outage, because the bowls were clean at that time. Dye penetrant exams were conducted on 2D which identified indications. The root cause of the leakage was determined to be PWSCC.

Also at Catawba Unit 2 on September 19, 2001, a walkdown of the SG 2B lower head bowl drain indicated boron residue buildup on the ½-inch piping immediately below the SG. The root cause of the SG 2B bowl drain leak was PWSCC of Alloy 600 material. The 2B SG bowl drain was repaired by plugging the drain, and tested satisfactorily. The remaining three SGs on Unit 2 were visually inspected and liquid penetrant tests were performed. No similar leaks were detected on the earlier repairs on the three other SGs.

In 2005, Calvert Cliffs Unit 2 found 2 axial cracks and 1 circumferential crack on the 2 inch drain line off the hot leg. The cracks were found during dissimilar metal weld exams as part of the plant's risk-informed inservice inspection program, and were repaired used weld overlay techniques. The weld overlay technique involves application of one or more layers of corrosion-resistant weld metal on the outside of the pipe, or penetration. This type of overlay provides both a less susceptible material, as well as a reduction of stress at the inside diameter.

2.4 Evaluation of the Susceptibility Model Using Effective Degradation Years

As explained previously, Alloy 600 cracking has progressed from one location in the plant to another. Along this line of cracking progression, VHP cracking has been one of the most significant issues. An example of the significance is the serious degradation of the Davis-Besse reactor head. One of the NRC actions resulting from the Davis-Besse event was the issuance of Order EA-03-009. This Order established interim inspections requirements for reactor pressure vessel heads depending upon their EDYs, using a formula specified in the Order.

When the susceptibility model was initially developed (Ref. 9), it was intended that several parameters could be factored into the formula. In addition to time and temperature, stress was incorporated into the original formulation. The fourth power of the stress was selected as a factor in the formula, based on Bandy and van Rooyen (Ref. 9) studies of crack initiation in cold-worked nickel-based alloys. Other parameters were considered, including hardness, yield strength, and carbide coverage of the grain boundaries. Initially, the susceptibility expression took the form of an expression for time, normalized to a reference time, and contained stress, referenced to a reference stress, and a materials susceptibility factor, K, similarly referenced.

$$t = t_{ref} \left[K_{ref} \middle/ K \right] \left[\sigma_{ref} \middle/ \sigma \right]^4 \exp \left\{ Q \middle/ R \left[1 \middle/ T - 1 \middle/ T_{ref} \right] \right\} \tag{1}$$

The difficulty of unequivocally computing stress for individual heads or nozzles soon led to elimination of stress from the equation. Similarly, lack of appropriate description of the microstructure of the Alloy 600 nozzles, exacerbated by confusing variations in the results of laboratory testing that addressed microstructural effects, led to elimination of the materials susceptibility factor from the equation. This left only time and temperature in the index, leading to the formulation used at the present time.

In the intervening years since the initial presentation of the susceptibility model, there has been a substantial increase in the amount of laboratory data and plant experience, that both support the model's fundamentals, while at the same time suggesting that there may be some possibility of improvement.

The susceptibility, or time-at-temperature model being used to compute an effective description of aging of nuclear steam supply system (NSSS) components is fundamentally based on the Arrhenius, or thermal activation model in which the rate of a single process, P, is a function of temperature (T, in degrees Rankin) and an activation energy (Q), expressed as:

$$P \propto \exp(-Q/RT) \tag{2}$$

where $\propto$ is the proportionality symbol and R is the universal gas constant (1.103×10^{-3} kcal/mole-°R). This formula has been used to describe a large number of kinetic processes that are dependent on thermal activation, including material deformation processes resulting from dislocation motion (Ref. 31). The initial application of this formula to vessel head penetrations may be attributable to Scott. In the very idealistic sense, "Q" should represent only one process, with an unchanging dependence on temperature. Consequently, the formula should not be extended to characterize a phenomenon which is dependent on multiple processes, each perhaps with a different dependence on temperature, or used over a temperature range so large that successive processes take place in specific ranges of temperature (oxidation of iron is a familiar example). Such misapplication of this formula would result in calculation of a Q-value integrated over the multiple processes. Pragmatically, however, a single expression may even be used for complex, multi-process phenomena, as long as the temperature range is relatively small, perhaps a few tens of degrees Kelvin. A small temperature range is the case for reactor heads, for which T ranges from about 560 °F to about 605 °F (~566 °K to ~591 °K).

The computation of EDY for a particular reactor component is based on the years of full-power operation (EFPY) normalized to 600 °F by incorporating the activation energy expression (Eq. 2) to achieve:

$$EDY_{600\,F} = \sum_{j=1}^{n} \left\{ \Delta EFPY_j \; \exp\left[-\frac{Q_j}{R}\left(\frac{1}{T_{head,j}} - \frac{1}{T_{600\,F}} \right) \right] \right\} \tag{3}$$

for applications to "PWSCC degradation of a reactor head," Q_j = 51 kcal/mole — the activation energy for crack initiation for Alloy 600. The summation over the index, j, allows for periods of time during which the component may have been subjected to different temperatures.[3] For example, this would apply to the head of a unit that had started up with one set of operating conditions (e.g., a relatively high head temperature), and was backfitted at some subsequent outage to provide cold leg flow diversion toward the head. Table 2-2 of MRP-48 (Ref 32) lists the vessel head temperature history for all domestic plants. NRC Inspection Manual Change Notice 02-037, dated October 18, 2002, contains explicit information on the calculations of EDY.[4] Also, it should be noted that the EDY susceptibility model is only applicable to the upper reactor vessel head and not the lower head (i.e., the lower head is at a cooler temperature) or other locations where Alloy 600 is used.

The Order requires EDYs to be determined for domestic power plants relative to 600 °F, or ~589 °K. The industry document describing crack growth rates (CGRs) of Alloy 600 (Ref. 5) also uses this same Arrhenius formula to put the CGRs on a parallel footing, but the reference temperature used is 617 °F (325 °C or 598 °K), and the activation energy used is 31 kcal/mole — the activation energy for crack growth in Alloy 600.[5]

---

[3]   Q has the same value, regardless of the temperature.

[4]   Manual Change Notice 02-037, NRC Inspection Manual, U.S. Nuclear Regulatory Commission, Washington, DC, October, 18, 2002.

[5]   None of that discussion pertains to Alloy 82/182 weld metal. The research toward determination of activation energies for either crack nucleation or growth in Alloy 182 or 82 is insufficient at the present time. Limited available research suggests that the activation energy for crack growth in Alloy 182 is about 125 percent of that of Alloy 600.

As described previously for Order EA-03-009, the plants were divided into four susceptibility categories: high, moderate, low, and replaced. High-susceptibility plants are those with an EDY greater than 12 years. Moderate-susceptibility plants are those that have an EDY between 8 and 12 years. Plants classified in the low-susceptibility category have an EDY less than 8 years. Lastly, plants in the replaced category are those that have replaced the RPV head and (therefore, by definition) have an EDY less than 8 years and have not experienced PWSCC in nozzle penetrations or the J-groove weld (Ref. 27).

Using data supplied to the NRC by the Materials Reliability Program (MRP), Figure 3 is a plot of the EDY calculations for domestic plants, using firm operational data at February 28, 2001, and an approximation to update all values to December 31, 2002. The symbols filled with red designate plants that have discovered VHPs with cracks. This plot also shows an approximated calculation of the EDY ($\approx$ 11.9) for Davis-Besse in February 1996, which is the earliest that Nozzle #3 is suspected to have begun to leak, according to the root cause report (Ref. 33). The plot shows that most plants either finding leaks, or making repairs, are well into the high-susceptibility range. The exceptions are the D.C. Cook 2 plant, which repaired one crack at EDY = 9.5 and has been free of leaks since (now at EDY = 13.9), and the Millstone 2 plant, which repaired non-leaking cracks in three nozzles at EDY = 11.6, after experiencing a clean NDE exam at EDY = 10.1.

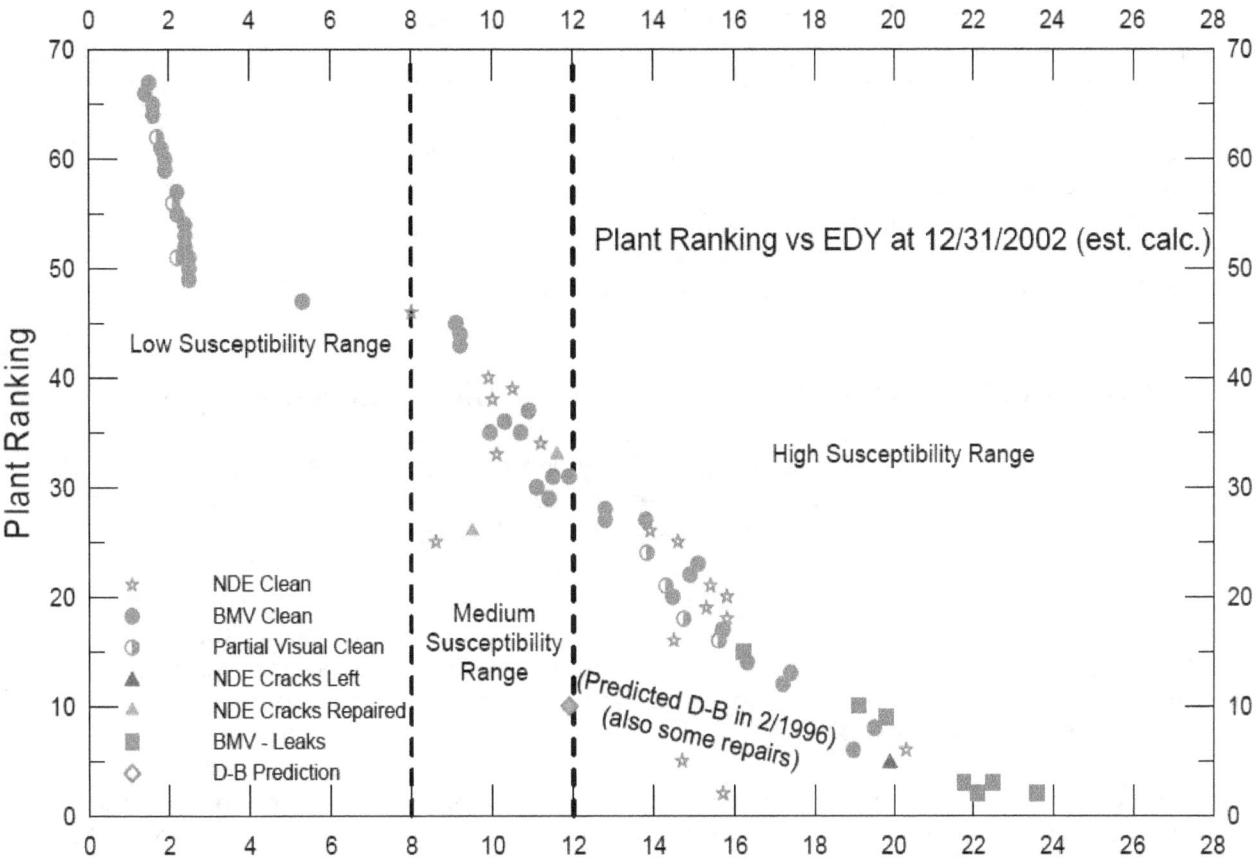

Figure 3. Ranking of domestic plants according to the EDY formula,
showing results of inspections, evidence of leakage, and repairs.
Many plants are shown with multiple symbols, indicating a "clean" inspection
at inspection opportunity, followed by a different finding at a subsequent inspection
(e.g., Oconee 2: clean NDE @ EDY=15.7, leaks and circ. flaws @ 22.1)

The experiences at the Millstone and D.C. Cook plants, and possibly at Davis-Besse, challenge the choice of 12 EDY as the medium-to-high-susceptibility threshold. However, there are several plants with EDY > 12 and clean inspection results. For example, examination of the nozzles in the H.B. Robinson 2 RPV head did not identify evidence of leaking or cracking. This suggests that there may be other factors besides time and temperature that are controlling crack initiation in VHP nozzles. One of these factors may be the heat of the material in the head. If variables, such as the heat of a material in a reactor head, can cause a contribution to the EDY calculations, such variables are not factored into the susceptibility model in its present incarnation. However, because there was an urgency to develop a tool to measure a plant's EDY after the Davis-Besse event, the current time-at-temperature EDY measurement tool was used. This was the logical choice because the time-at-temperature EDY calculation reasonably reflected industry cracking data, while leaning toward a more conservative position.

VHP nozzle inspection results indicate that the cracking susceptibility approach provided in Order EA-03-009 is effective in designating groups of plants for specific types of inspections, and do not indicate a need for revising the EDY model in the short term. However, the actual EDY calculation is becoming less of an issue as a result of reactor vessel head replacements. The cost of 100-percent volumetric inspection of heads, which is required at every outage for the high-susceptibility plant, dictates that replacement of the head is an effective way to manage VHP nozzle cracking. As of March 2005, 14 U.S. plants have replaced their reactor heads. With many of the high-susceptibility plants replacing their heads with new Alloy 690 heads, many plants will fall into the low-EDY susceptibility category.

2.5   Current Research on Degradation of Nickel-Base Alloys

Because the Alloy 600 cracking issue is still not fully characterized, there are many research projects in progress, both domestically and internationally. The Office of Nuclear Regulatory Research (RES) programs covering the scope of flaw evaluation for light-water reactor applications are characterized by studies of (1) nondestructive inspection technologies, (2) materials properties and crack growth rates, and (3) stress and structural integrity analysis. These three diverse technologies must be tied together in order to provide an accurate understanding of the crack nucleation and growth phenomenon and, from that, to predict appropriate intervals for inspection, or in some cases, computation of times to leakage (i.e., for a crack to go through-wall).

Crack growth rate testing in simulated reactor coolant environments is a very challenging and time-consuming procedure. Measurement of a single crack growth rate data point in the 10^{-11} to 10^{-9} m/s range requires several weeks of uninterrupted high-temperature, high-pressure exposure in simulated PWR or BWR coolant. During this time, load on the specimen must be held constant, and the environmental chemistry must be carefully controlled. This environmental control must be coupled with extremely precise crack extension measurement instrumentation, capable of measuring 0.02 mm of crack extension, precisely, and without any drift or systematic error. Also, such tests must begin with a conditioning (sometimes called "coaxing") phase, usually of about three weeks, in order to achieve equilibrium of the electrochemical environment, and establish the desired mechanism for crack extension.

With regard to Alloy 182/82, far less crack growth rate data are available, and the dependence of the data on variables such as temperature, time, stress, etc., is not well known. There is, however, general consensus that crack growth rates are about a factor of five faster in Alloy 182 than in Alloy 600. Test programs for thick section Alloy 690 and its companion weld metals, Alloy 152/52 are virtually nonexistent, but several are slated to begin in 2005. Some additional work is being undertaken by vendors and owners' groups, notably some Alloy 600, 690, and associated weld materials crack growth rate testing underway in the Westinghouse Science and Technology Center.

Under the present terms of the inspection orders and bulletins, most licensees with plants deemed as moderately or highly susceptible to SCC have made the decision to replace their reactor vessel heads. Therefore, flaw evaluations for the long term (for vessel head penetrations) are a somewhat academic exercise for those plants at this time. However, flaw evaluations for other components (safe-ends, small bore nozzles and piping, as examples) are a reasonable possibility. Evaluation of flaws in replacement heads, usually fabricated with nozzles and welds of Alloy 690 (for the VHPs) and Alloy 152 (for the attachment welds) is an even more likely possibility, since the crack growth rates are considerably less than those for the corresponding Alloy 600 and Alloy 182 used in the original heads.

The last aspect of flaw evaluation that should be mentioned in this context concerns the mitigation of flaw growth. There are several approaches to slowing the nucleation and growth of cracks, including temperature reduction, chemical additions (usually zinc), and careful control of the electrochemical potential (usually by maintaining hydrazine additions at the high side of the water chemistry guideline). Vessel upper head temperature reduction (UHTR), managed by diverting some of the cold leg return flow into the upper dome, has been applied to a number of U.S. and non-domestic vessels. Several U.S. licensees are adding zinc to the primary water, in quantities in the range of 10 to 20 parts per billion, which is allowed within the EPRI water chemistry guidelines, and most plant technical specifications.

While there is substantial work being conducted in the U.S., foreign countries are also participating in the research activities related to Alloy 600. In Japan, vessel heads with Alloy 690 penetrations have replaced the original heads at 12 of the 23 PWRs. T-cold conversion (i.e., UHTR) has been implemented at the remaining 11 plants. Of these 11, 6 have undergone eddy current inspections within the past 10 years, and no flaw indications were found. Eddy current inspection and water-jet peening of the BMI nozzles in seven plants has been completed at the present time. This technology is being developed for VHP penetrations. Water-jet peening technology is also being developed for safe-end nozzles. Two, large nickel-base alloy SCC programs are underway. The Electric Joint Research Project includes tasks to evaluate SCC and slow strain rate testing (SSRT) on Alloy 600, Alloy 132, 82, thermally treated (TT) 690, and Alloys 152 and 52. The National Nickel-Based Alloy Material Project, sponsored by a consortium of Japanese utilities, vendors, and government agencies, includes tasks to evaluate SCC on mill-annealed Alloy 600 (600 MA), Alloy 132, 82, thermally-treated Alloy 690 (TT690), and Alloys 152 and 52. Both of these programs are long-term, with ending dates in the 2005/2006 time frame.

In France, the discovery of VHP cracks in 1991 led to analysis of the effects of temperature and water chemistry, the effects of stress, and the factors behind the susceptibility of nickel-base alloys to SCC. However, the early decision to replace heads allowed the French to focus on repair strategies for the relatively short term, together with inspection programs for the heads that had not been replaced. Since the inspection policy is crack growth rate dependent, the inspection targets are (1) every 3 years for cracks less than 3 mm deep, (2) every 2 years for cracks between 3 and 5 mm deep, and (3) every year for cracks above 5 mm. In all cases, the head would be replaced before reaching a safety criterion of 4 mm of remaining ligament for the 900 MWe reactors. Other penetrations are also inspected, on a sampling basis, and include the steam generator partition plate stub welds, BMIs and earlier nozzle repairs. Twenty-six steam generator (SG) divider plates, determined to fall into a high-susceptibility category ("precursor" is the French terminology), will be inspected during the 2001–2008 time frame. Nine additional, randomly chosen SG divider plates will be inspected. The 10-year inspections have been completed at several (< 5) of the French plants with replaced heads, and no indications of cracking were discovered. Of 54 heads, 42 have been replaced and 2 more replacements were scheduled for 2004.

3.0 BORIC ACID CORROSION
OF LIGHT-WATER REACTOR PRESSURE VESSEL MATERIALS

Boric acid leakage is a consequence of Alloy 600 cracking. This leakage can lead to boric acid corrosion of low-alloy steel that is in contact with the aqueous solution or dried solution products. The degradation of the Davis-Besse reactor vessel head is an example of the severe consequences of boric acid corrosion.

3.1   Introduction

Boric acid corrosion, or wastage presents a significant maintenance problem for PWRs, which use borated water to control reactivity during normal plant operations. Boric acid can severely degrade low-alloy and carbon steel under the right conditions (Ref. 34). While many research programs have been conducted to study the effects of boric acid corrosion of reactor components, these studies fall short in many areas. Little work has been done on the corrosion of low-alloy and carbon steels in (1) molten salts of boric acid, (2) high-pressure and temperature boric acid containing environments, or (3) aqueous solutions of boric acid in the temperature range of 100–250 °C (Ref. 35). This report will outline the NRC and industry actions with respect to boric acid corrosion inspection, summarize the pertinent foreign and domestic boric acid corrosion events, and review the most recent and ongoing boric acid corrosion test programs.

Over the past 35 years, many incidents of boric acid corrosion have been observed on PWR components. The majority of reported boric acid corrosion events involved reactor coolant that leaked from flanged joints and corroded the threaded fasteners of primary valves and pumps, or the threaded closure studs on manways. Events involving cracked and leaking RPV head penetrations leading to corrosive attack of the vessel head were reported with less frequency.

3.2   Findings at Davis-Besse

Degradation of the RPV head at Davis-Besse was discovered in March 2002. The integrity of the reactor coolant boundary was significantly compromised. A triangular cavity approximately 5 inches wide, 7 inches long, and completely through the low-alloy steel RPV head thickness was located downhill of VHP Nozzle #3 approaching Nozzle #11. The wastage consumed between 40 and 60 cubic inches of the vessel head. This reduced the reactor pressure boundary in the vicinity of the boric acid corrosion attack to a section of stainless steel cladding ranging in thickness from 0.199 inch to 0.314 inch. The cladding had an exposed surface area of 16.5 square inches and had deflected outward under system pressure. Additional boric acid wastage was also observed near VHP Nozzle #2. Figure 4 is a photograph of the extensive corrosion loss experienced on the head at the Davis-Besse plant.

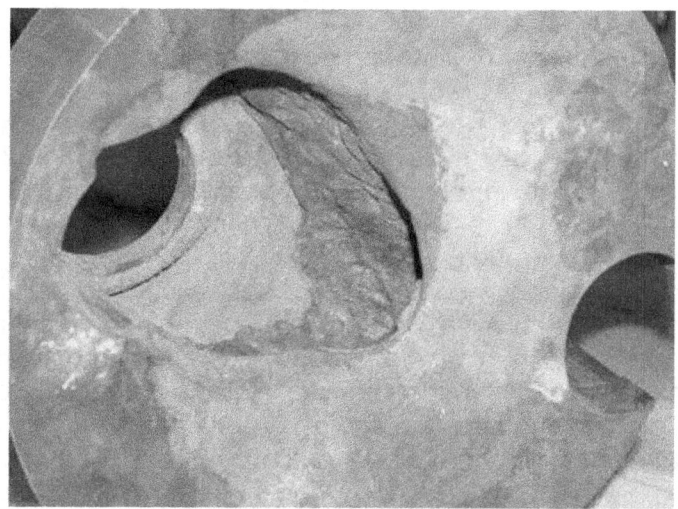

Figure 4. A photograph of the 18-inch diameter section of the discarded Davis-Besse head, containing the corrosion cavity created by leakage from the included nozzle (#3). The hole at the right is from nozzle #11.

Prior to the Davis-Besse incident, the total amount of leakage through any of the through-wall cracks in a VHP nozzle into the annulus at any other plant was low, and occurred at low leakage rates. In the case of Davis-Besse, evidence of a leak was not observed at an early stage of its occurrence. The leak rate escalated as the axial crack extended resulting in significant RPV head corrosion. The exact mechanisms that led to this corrosion are not fully understood. However, at the time of the event a root cause report was issued describing the industry's understanding of the contributing factors that lead to the event (Ref. 33)

3.3 NRC Actions To Address Boric Acid Corrosion Issues (1980–2004)

In May of 1980, Omaha Public Power District informed the NRC staff that severe corrosion damage had been found on a number of closure studs in two of four reactor coolant pumps (RCPs) at Fort Calhoun Unit 1. Three of the four RCPs at Fort Calhoun exhibited primary water leakage at the pump cover/case interface. After removal of the nonmetallic insulation, further inspection revealed three studs located side-by-side on one pump and three studs similarly located on another pump had significant boric acid corrosion. The 3.5-inch diameter studs securing the RCP covers are made from American Society for Testing and Materials (ASTM) A193 Grade B7 low-alloy steel with chrome-plated threads. The studs experienced severe boric acid corrosion wastage with a maximum reduction in diameter of 1.1 inch. In June of 1980, the NRC issued IN 80-27 concerning the failure of low-alloy steel primary coolant pump cover studs as a result of boric acid corrosion. The notice also discussed the deficiency of current ultrasonic testing (UT) procedures in revealing stud wastage and stressed the need for supplemental visual inspections (Ref. 36). In 1982, the NRC issued two generic communications related to boric acid corrosion of carbon and low-alloy steels. IN 82-06, issued in March 1982, discussed the failure of steam generator primary manway studs at Maine Yankee. The studs, which failed as a result of stress-corrosion cracking, also showed indications of surface corrosion attack resulting from the interaction of stud preload, primary coolant, lubricants, and Furmanite sealant (Ref. 37).

Bulletin 82-02, issued in June of 1982, was a followup to INs 80-27 and 82-06. The Bulletin notified licensees of incidents of severe material degradation of threaded fasteners at Fort Calhoun and Maine Yankee. At Maine Yankee, the 1-inch diameter steam generator primary manway studs made of SA 540 Grade B 24 alloy steel failed by stress-corrosion cracking. Bulletin 82-02 discussed the agency's concern about accelerated corrosion and stress-corrosion cracking as a result of chlorine, fluorine, and sulfur contamination from sealants and lubricants used in the reactor coolant pressure boundary (RCPB) and suggested that care be taken when choosing sealants and lubricants for this system. Expanding on the discussion in IN 82-06, the bulletin highlighted the inadequacy of UT to detect stress-corrosion cracking and wastage of threaded fasteners. To correct the shortcomings of the American Society of Mechanical Engineers (ASME) Code UT procedures, the agency suggested a combination of NDE techniques, including UT, visual examination (VT-1), dye penetrant examination (PT), and magnetic-particle testing (MT). Licensees were required to take specific action with respect to inspection and maintenance of threaded fasteners susceptible to boric acid leakage in the RCPB. Additional information was requested in regard to the licensees' operational experience with bolted closures, sealants, and lubricants used in the RCPB. Concerns raised with regards to the UT procedures current at that time led to the 1983 revision of the ASME Code Section XI to include more disciplined and wide-ranging requirements for visual examination of systems containing primary coolant (Ref. 38).

In response to the continued occurrence of ferritic component boric acid degradation, the NRC issued IN 86-108 (Ref. 39) and its supplements (Refs. 40 to 42). Notice 86-108 informed licensees of the severe boric acid wastage of the carbon and low-alloy steel components of the high-pressure injection (HPI) nozzle and reactor coolant system cold leg pipe at ANO-1. A 6-month-long leak from an overhead high-pressure injection (HPI) valve caused boric acid wastage of a maximum depth of 0.5 inch and 0.25 inch on the underside of the HPI nozzle and the adjacent RCS piping, respectively. IN 86-108 references the 1984 EPRI work, entitled EPRI-NP-3784, "A Survey of the Literature on Low-Alloy Steel Fastener Corrosion in PWR Power Plants" (Ref. 43). This report describes the Combustion Engineering laboratory tests in which borated water was dripped onto a hot metal surface, similar to the ANO-1 experience. The report concluded that aggressive boric acid corrosion of locally-cooled low-alloy steel parts can occur when the plant is at operating temperatures. The dripping borated water solution concentrates as the water boils off. The evaporation process locally cools the metal surface to the boiling point of concentrated boric acid solution, the temperature at which boric acid corrosion is at a maximum (Ref. 39).

Supplement 1 of IN 86-108 was issued in March of 1987. IN 86-108, Supplement 1 described the severe boric acid wastage of the RPV head at Turkey Point Unit 4. Borated water from a conoseal joint in an instrumentation port column assembly leaked onto the RPV head, near the edge of the vessel head. The leak, discovered during an August 1986 outage, was determined to have a low probability of causing significant corrosion wastage. During the March 1987 outage, however, approximately 500 pounds of boric acid crystals were discovered covering part of the RPV head. The vessel head flange and several flange nuts and studs were severely corroded. The most extensive corrosion was of the vessel head in the form of a "boomerang-shaped" depression, with dimensions of 8.5 inches long x 1.25 inch wide x 0.25 inch deep (Ref. 40).

The NRC issued Supplement 2 to IN 86-108 in November 1987. The IN informed licensees of pinhole leaks in the seal weld of the conoseal for thermocouple connections to the RPV head at Salem Unit 2, which resulted in RPV head degradation. The RPV head corrosion consisted of nine corrosion pits that were 1–3 inches in diameter and 0.36–0.40 inch deep (Ref. 41).

In response to the events in IN 86-108, including Supplements 1 and 2, Westinghouse performed experiments on the corrosion effects of primary coolant leaking on low-alloy and carbon steels and concluded that boric acid corrosion rates are greater than those previously known or estimated in other studies. Westinghouse conducted corrosion tests of carbon steel in 25 percent boric acid at 200 °F, which produced corrosion rates of 400 mils/month for aerated solutions and 250 mils/month for deaerated solutions. In addition, Westinghouse created a VHP head weld mock-up that had a typical crevice geometry. The RPV head weld was exposed to dripping 15 percent boric acid at 210 °F. The results of this experiment were extensive corrosion (400 mils/month) of the carbon steel vessel head and virtually no wastage of the Inconel or weld metal (Ref. 41). This type of behavior is similar to that seen at Davis-Besse in 2002.

The NRC issued Generic Letter (GL) 88-05 in March 1988 to address the issue of boric acid corrosion of low-alloy and carbon steels in the reactor coolant system (RCS) as a result of RCS leakage at less than the limits specified by the technical specifications (TS). The GL required licensees to develop a program of inspection, leak identification, engineering evaluations and corrective actions. More specifically, it requested licensees to identify locations where degradation can occur as a result of RCS leakage rates less than the TS limits. GL 88-05 required licensees to determine the leak paths and the components that would potentially be affected by boric acid corrosion. GL 88-05 required licensees to develop methods of inspection to identify and locate these leaks, methods of performing engineering evaluations to determine the risk of degradation once the leak is identified, and a system of corrective actions to preclude recurrence of boric acid corrosion (Ref. 44).

In January of 1995, the third and final supplement to IN 86-108 was issued. It described the 1994 boric acid corrosion events at Calvert Cliffs Unit 1 and Three Mile Island. At Calvert Cliffs Unit 1, boric acid leaking past a flange gasket corroded three nuts on an incore instrumentation flange. Maintenance personnel at Three Mile Island found that four out of the eight total studs holding the pressurizer spray valve bonnet gasket in place had corroded. At both plants, as with the Turkey Point Unit 4 event, the boric acid leakage was identified in a previous outage and the risk of boric acid corrosion was determined to be low. The IN concluded that this pattern of behavior may be indicative of an underlying lack of awareness of the conditions and mechanisms that lead to boric acid corrosion (Ref. 42).

GL 97-01 was issued in 1997 to address the problem of VHP cracking at the dissimilar metal vessel head weld that joins the VHP housing to the reactor vessel head, and requested that licensees perform inspections of these welds. In addition, the letter described the action plan implemented by the NRC in 1991, which included a review of the owners groups' safety assessments, EPRI's RPV head penetration mock-up tests, review of NEI's proposed generic acceptance criteria, and vessel head penetration inspections (Ref. 21). The owners' groups' assessments, specifically the Babcock & Wilcox Owners Group, had the foresight to model a leaking VHP nozzle penetration and its consequences (Ref. 45). The model was based on the possibility of severe boric acid wastage of the RPV head with a corrosion rate of 16.7 cm^3/yr (1.07 in^3/yr), however, the responses of the owners' groups demonstrated that the structural integrity of the head (in terms of the ability of the head to sustain the internal pressure of the coolant) would not be degraded as a result of a corrosion loss of this extent. Data presented in a later section shows that much higher rates of corrosion are plausible under certain conditions of temperature and concentration. A structural integrity calculation based on the higher corrosion rates, and a consequent greater wastage of low-alloy steel, would demonstrate a much reduced margin of safety.

As described previously, while conducting inspections required in Bulletin 2001-01, severe boric acid corrosion wastage of the reactor pressure vessel head was discovered at Davis-Besse Nuclear Power Station in February 2002. In response to this event, the NRC issued Bulletin 2002-01, which required licensees to (1) report on current inspection and maintenance practices of the reactor pressure vessel head and assess their effectiveness, (2) submit a summary of their boric acid inspection programs (per GL 88-05), and (3) give information on the material condition of the reactor pressure vessel head and the remainder of the RCS pressure boundary (Ref. 46). The licensees' responses to this bulletin led to the publication of Regulatory Issue Summary 2003-013, "NRC Review of Responses to Bulletin 2002-01, 'Reactor Pressure Vessel Head Degradation and Reactor Coolant Pressure Boundary Integrity' (Ref. 47)." The NRC noted in Regulatory Issue Summary (RIS) 2003-13 that most licensees do not perform inspections of Alloy 600/82/182 materials in addition to those required by Section XI of the ASME Code to identify potential cracked and leaking components. RIS 2003-13 also stated that it is the NRC staff's understanding that many licensees perform the ASME Code-required inspection without removing insulation from around the head and penetrations and, therefore, may not be able to detect the amounts of through-wall leakage expected from potential flaws attributable to PWSCC or other cracking mechanisms.

IN 2003-02, issued in January 2003, addressed the boric acid corrosion of the reactor pressure vessel head at Sequoyah Unit 2 and the VHP housing leak at Comanche Peak Unit 1 (Ref. 48). Both Bulletin 2002-01 and IN 2003-02 contest the assumption that primary water leakage onto hot surfaces will not result in boric acid corrosion. The recent findings at Davis-Besse (2002) and Sequoyah Unit 2 (2003) have challenged these assumptions and led to increased experimentation into the mechanisms and conditions that produce boric acid corrosion.

As previously described, NRC Order EA-03-009, issued in February 2003 and revised in February 2004, strengthened PWR inspection plans with respect to the reactor pressure vessel head and penetration nozzles (Ref. 26). The Order requires in part that visual inspections to be performed during each refueling outage in order to identify potential boric acid leaks from pressure-retaining components above the RPV head. For plants with boron deposits on the surface of the RPV head or related insulation (discovered either during an inspection required by this Order or otherwise and regardless of the source of the boron), the licensee is required to perform inspections of the affected RPV head surface and penetrations to verify their integrity before returning the plant to operation.

As a result of numerous requests to relax the previous orders, the Commission issued the first revised Order EA-03-009 in January 2004. The 2004 revision to Order EA-03-009 specifies that if the surface of the RPV head located downslope of the outermost RPV head penetration is obscured by support structures, the BMV inspection must include no less than 95 percent of the vessel head and, if any corrosion products or evidence of boron is detected, the support structure must be removed and the obscured region must be inspected.

3.4   Foreign and Domestic Events

The first reported incident of boric acid corrosion occurred at Haddam Neck in 1968. Carbon steel valve bonnet bolts were severely degraded after a short exposure to a primary water leak. This event identified the need to avoid exposure of low-alloy and carbon steels to borated water and prompted their replacement with more corrosion resistant materials in applications that did not require the use of high-strength materials (Ref. 43).

Following this first reported U.S. incident, several incidents of boric acid corrosion of threaded fasteners were reported during the 1970s, 1980s, and 1990s including events at Palisades, Zion Unit 1, Calvert Cliffs Units 1 and 2, St. Lucie Unit 1, Surry Unit 2, ANO-1 and ANO-2, Maine Yankee, Fort Calhoun, HB Robinson, ONS-2 and ONS-3, Millstone Unit 2, Indian Point Unit 2, DC Cook Unit 2, Kewaunee, North Anna Unit 1, San Onofre Unit 2, Waterford, and Davis-Besse (Refs. 34, 43, 49, and 50).

In 1970, 1 year after starting operation, the Swiss reactor Beznau Unit 1 experienced some inconsequential boric acid wastage of the low-Alloy steel RPV head. A canopy seal leaked at a weld defect and produced a large deposit of boric acid crystals at the RPV upper head. After removing the deposit and cleaning the head, a crescent-shaped region of attack was observed adjacent to the CRDM nozzle penetration. The corroded region was 50 mm wide and 40 mm deep. Dye penetrant testing and stress analysis were performed on the vessel head before it was returned to service without repair (Ref. 51).

Boric acid corrosion of a weld in the suction piping of the reactor coolant pump was discovered during an inspection at Calvert Cliffs in 1981. The dissimilar weld joint was composed of a stainless steel clad carbon steel elbow that was welded with Inconel Alloy to a stainless steel safe end. The wastage, located on the surface of the carbon steel near the dissimilar metal weld, extended 18 inches circumferentially around the 30-inch outside diameter pipe. The region of corrosion attack penetrated • •inch into the 3½-inch wall thickness (Refs. 43 and 50).

Also in 1981, boric acid corrosion of an instrument isolation valve bonnet was observed at Kewaunee. The bonnet assembly was replaced within the TS-allowed outage time, negating the need for a plant shutdown (Refs. 50 and 52).

In 1987, PWSCC of a pressurizer heater sleeve, made of Alloy 600, resulted in the leakage of primary water onto the pressurizer head at ANO-2. The low-Alloy steel (SA-533 Grade B Class 1) head experienced boric acid wastage that was 1½ inch in diameter and ¾ inch deep (approximately 18 percent through-wall). The leaking nozzle was plugged and the pressurizer vessel head was weld repaired before being returned to service (Ref. 50).

In 1988, at Millstone Unit 2, the RPV O-ring seals were found to be leaking on three occasions, one of which caused boric acid corrosion of the closure region. The wastage encompassed nine RPV studs and two small regions of the cold leg nozzles. The RPV studs were replaced and the cold leg nozzles were cleaned and returned to service (Ref. 34).

Boric acid corrosion of the containment liner, which spanned 30 feet in length and 0.1 inch of the 1 inch original depth, was discovered at McGuire Unit 2 in 1989. Borated water from a leaking instrument line compression fitting pooled on the containment liner and led to boric acid corrosion attack (Ref. 34).

In 1989, the nut ring and bolts from beneath the VHP nozzle flange at location L-2 at ANO-1 were degraded as a result of boric acid corrosion. Approximately 50 percent of one of the nut ring halves had corroded away and two of the four bolt holes in the corroded nut half-ring were degraded such that there was no bolt/thread engagement. An inspection of the flanges and the spiral wound gaskets which were removed from between the flanges revealed that the cause of the leaks was the gradual deterioration of the gasket material with age. The gasket at L-2 had been in place since initial plant operation. The initial leakage flowpath at L-2 was across the face of the flange, down the hold down bolt counterbore, then contacting the carbon steel nut ring. This leakage flowpath had not previously been anticipated. Because the leakage did not travel to the outside edge of the flange and because the nut ring area is not easily viewed, it was not detected during previous routine inspections of the RPV head area (Ref. 53).

During the period between 1989 and 1994 several leaks resulting in wastage with a maximum depth of ½ inch were found in the carbon steel reactor coolant charging pump casings at North Anna Units 1 & 2. The wastage was associated with cracks in the stainless steel cladding (Ref. 54).

In 1991, the French reactor Bugey 3 experienced a leak from a VHP nozzle penetration into the RPV head. Water and boric acid crystals were found at the base of the penetration and a small circumferential crack was observed at the root of the J-groove weld. Over a period from 1991 through 1996, extensive destructive examination was performed on the penetration. To better understand the effect of the annulus leak on the low-alloy steel vessel head, the area of the vessel head that made up the annulus was "peeled" using six transverse cuts. The surface was examined and a leak path was identified. Boric acid corrosion wastage of a maximum depth of 60 µm and a duplex corrosion product with a maximum thickness of 65 µm was associated with the leak path. This event led to the 1993 decision by Electricité de France to replace all French RPV heads that had Alloy 600 penetrations (Ref. 9).

In 1994, Calvert Cliffs Unit 1 experienced higher than expected corrosion rates of three carbon steel nuts and one incore instrumentation (ICI) flange on the RPV head. The boric acid corrosion was attributable to reactor coolant leakage past the ICI detector joint assembly, which caused the build-up of concentrated boric acid on the flange components. Repairs of the leaking flanges, which were known to have been leaking since 1993, were deferred until 1994 since the licensee assumed that the components would be exposed only to dry boric acid crystals and, thus, the corrosion rates were expected to be low. The licensee's assertion that the boric acid would be dry was based on an incorrect estimated flange temperature of 500 °F, which would have been high enough to boil off the moisture in the boric acid. The actual flange temperatures were later measured as between 160 °F and 295 °F, in the temperature range of the greatest boric acid corrosion rate (Ref. 55).

In 1996, workers in a Westinghouse shop discovered blistering, pitting, and linear indications in a Callaway main coolant pump. These flaws were located outside of the gasket contact area of the two opposing flanges of both the #1 seal and the thermal barrier. The blisters and indications could be detected both visually and tactilely, and there was evidence of minor boric acid wastage on the threaded area of the carbon steel studs. Crystalized boric acid was discovered between the two flanges that were made of SA-182 Grade F 304 stainless steel. The boric acid crystals had become wetted, and resulted in the continued boric acid corrosion of the studs, which was evidenced by the blistering and pitting (Ref. 34).

In 1996, the French reactor Bugey 3 experienced further boric acid attack. A leak originated from a misaligned bolted flange on a pressure vessel air vent line. Tricastin 4, another French reactor, experienced a leaking canopy seal in 1998. Both of these events produced significant boric acid deposits; however, the depth of boric acid attack to the upper vessel head was only a few millimeters and did not require repairs (Ref. 51).

In 1999, at Davis-Besse, the valve packing of nine primary pressure boundary valves leaked and resulted in the boric acid corrosion of the carbon steel yokes. The most severe case involved the wastage of 80 percent of the yoke's cross-section. The valves were repaired and inspected for zero leakage before being returned to service (Ref. 34).

In 2000, at VC Summer, a through-wall crack in an Alloy 182 weld between a carbon steel hot leg and a stainless steel pipe resulted in more than 200 pounds of boric acid crystals accumulating near the welded joint. When the joint was cleaned, visible evidence of boric acid corrosion was discovered on the carbon steel side of the joint, however, the depth of wastage was immeasurable (Ref. 34).

The event that has attracted the most attention to date is the degradation of the RPV head at Davis-Besse which was discovered in March 2002. As described in Section 3.2 and Refs. 1 and 33, the integrity of the reactor coolant boundary was significantly compromised. A triangular cavity approximately 5 inches wide, 7 inches long, and completely through the low-alloy steel RPV head thickness was located downhill of VHP Nozzle #3 approaching Nozzle #11. Additional boric acid wastage was also observed near VHP Nozzle #2.

Sequoyah Unit 2 also experienced head wastage in 2002, yet to a much lesser extent than at Davis-Besse. During a forced outage, the licensee discovered a primary coolant leak from an improperly reassembled compression fitting on the reactor vessel level indication system (RVLIS). The leaking fitting sprayed a fine mist of reactor coolant at a rate of approximately 0.001 gpm onto the vessel head insulation. The coolant seeped through a seam in the insulation and onto the low-alloy steel vessel head, where it resulted in the wastage of a finger sized groove, which was 5 inches long, $5/_{16}$ inch wide, and $1/_{16}$ inch deep. During the previous outage, there were no indications of vessel head corrosion noted during a bare metal visual inspection; the forced outage and the observation of corrosion took place 7 months later (Ref. 56).

At the low leak rates experienced at Sequoyah, the coolant would be expected to boil off quickly when it came in contact with the hot vessel head, leaving dry boric acid crystals that cause extremely low corrosion. However, the continuing mist of coolant may have humidified or hydrated the boric acid crystals, which resulted in the observed corrosion rate of 0.02 inch/month (Ref. 56). The results of the NRC boric acid corrosion program that was performed at ANL (Ref. 35) will discuss this phenomenon in greater detail.

During the 2003 refueling outage at Three Mile Island, Unit 1 (TMI-1), inspectors discovered boric acid crystals between the pressurizer heater bundle (PHB) diaphragm plate and the PHB cover plate. The licensee determined that the leak path emanated from the lower PHB through the edge of the PHB diaphragm plate. During the previous outage, the licensee misdiagnosed the deposits as an inactive leak from a previous event. As a result of this oversight, the licensee did not remove the PHB cover plate to inspect for the leak path, and did not perform NDE to identify the flaw that led to the leakage. This omission allowed the degradation to proceed until the next outage in 2003 at which point it was determined that the leak had existed since 1998. The crack in the PHB diaphragm plate, caused by PWSCC, leaked borated water onto the carbon steel PHB cover plate and caused severe boric acid corrosion. The cover plate wastage was approximately 1.35 inch deep and 7 inches across. The licensee replaced the PHB, PHB diaphragm plate, and PHB cover plate to correct for this failure (Ref. 57).

During the same outage at TMI-1, leakage from a non-safety-related chemical addition valve, which was identified in the 2001 outage, resulted in the boric acid corrosion of the carbon steel containment liner. The wastage of the moisture barrier in the cylindrical portion of the containment liner spanned 20 feet in length, 2 to 4 inches in width, and had a maximum reduction in wall thickness of 18.4 percent. At the time that the leakage was identified in 2001, the licensee did not take into account the effect of the boric acid leakage, which was directly above the degraded area, on the containment liner and moisture barrier.

3.5   Current Research on Boric Acid Corrosion

The extensive corrosion of the low-alloy steel vessel head found at Davis-Besse in March 2002 created significant interest in discovering the set of conditions that led to the wastage. The initial failure analysis of that corrosion was completed in June 2003 (Ref. 33). To address the concerns raised by this issue, RES has initiated several programs. Pieces of the reactor vessel head were removed, and have been shipped to Pacific Northwest National Laboratory (PNNL), where they will be decontaminated, and the smaller corrosion cavity surrounding Nozzle #2 (the major corrosion was at Nozzle #3) will be examined in an attempt to discover the conditions near a (relatively) younger VHP leak. Other pieces of metal were removed from Nozzle #3, and the metal surrounding Nozzle #11, to be used for crack growth rate tests, characterization of the microstructure, and mechanistic studies based on crack tip analyses.

At ANL, an extensive program of corrosion testing of several pressure boundary materials in concentrated solutions of boric acid has been completed (Ref. 35). The output of this program will be data describing the corrosion rates and electrochemical properties of boric acid solutions over a range of temperatures, concentrations, and aeration levels.

The extensive cracking of Alloy 182 J-welds and VHPs at the North Anna 2 plant, including the observation of cracked VHPs that did not exhibit visible evidence of leakage, has provided the impetus for a large, collaborative program involving the NRC and EPRI/MRP. Before the head was disposed in Utah, six nozzles, with a substantial amount of the surrounding low-alloy steel and cladding, were cut from the head (at industry expense), and shipped to PNNL. The nozzles are being decontaminated, and will undergo nondestructive examination and sequential sectioning to determine qualitatively if the probability of detection and crack sizing was satisfactorily accurate.

While boric acid corrosion wastage is a known degradation mechanism, the severity of the RPV head wastage at Davis-Besse was not anticipated. Prior to the event, it was believed that at low leak rates typical of VHP nozzle through-wall cracking, approximately 10^{-6} to 10^{-5} gpm, the leaking coolant would completely vaporize to steam resulting in dry boric acid crystal deposits, which cause little or no corrosion. The Davis-Besse event proved that this is not always the case. In the case of Davis-Besse, the leak rate escalated as the axial crack extended resulting in significant RPV head wastage. The Davis-Besse Root Cause Report (Ref. 33) suggests a scenario of five steps that led this wastage:

(1)   *Crack initiation and growth to through-wall*: It is hypothesized in the root cause report that the crack initiated as a result of PWSCC in VHP Nozzle #3 in 1993. It is further hypothesized that between 1994 and 1996, the crack proceeded to grow through the J-groove weld, which attaches the VHP nozzle to the inside of the RPV head. During this stage, the extent of through-wall cracking was postulated to be extremely limited and the RCS leakage was thought to be small.[6]

(2)   *Minor weepage/latency period:* As the extent of through-wall cracking progressed axially, the RCS leakage would have entered the annulus, the region between the Alloy 600 VHP nozzle and the SA-533 Grade B low-alloy steel RPV head. The environment created in the annulus made possible several corrosion and concentration processes, including galvanic attack.

(3)   *Late latency period:* With further extension of through-wall cracking, the annular gap would presumably widen, and since the gap width extended over a considerable portion of the annular length at Davis-Besse, the annulus flow area would increase more quickly than the crack flow area. Annular plugging is ignored in the root cause analysis, favoring the opinion that the primary flow resistance would have been a result of the crack dimensions, and not of the restriction offered by the annular geometry. These conditions make possible the entrance of oxygen into the annulus, increasing wastage rates dramatically.

(4)   *Deep annulus corrosive attack:* In the scenario presented in the root cause report, the annulus would continue to widen, resulting in a decreased flow velocity out of the annulus, and a decreased differential pressure (D/P), which would in turn allow increased oxygen penetration and corrosion rates. The root cause report contends that corrosion is presumably the greatest in the vicinity of the crack since leakage through the crack would deliver new reactive oxidizing ions to the boundary layer of the corroding metallic surface.

(5)   *Boric acid corrosion:* Once the leakage escalated to a high rate, the annulus could have filled with an increasing amount of moist steam that would partially flash as it exited the annulus. The large heat of vaporization required to vaporize the leaking coolant would decrease the temperature in the steam in the leaked coolant, and locally suppress the metal surface temperature. Thus, heat transfer from the surrounding metal would no longer be adequate to immediately vaporize the remaining portion of the leakage that did not flash. This effect would allow for the increased wetting beneath the currently existing boric acid deposits. As the crack widened and the leak rate increased, the corroding annulus would begin to fill with increasingly concentrated boric acid solution. Since the wetted area is a result of liquid flow from the crack, it would be expected to be primarily downhill from the nozzle, which would result in the high corrosion and wastage rates of the material of this area of the RPV head.

At the time that the root cause report was written, there was not enough information to support the proposed sequence of events, nor the mechanisms that led to the observed RPV head

---

[6]   This version fo the progression of the corrosion came from References 33 and 35.

wastage. However, the degradation modes on the two extremes were well understood. For the extremely small leak rates observed in most leaking VHP nozzles, on the order of 10^{-6} to 10^{-5} gpm, the leakage will completely vaporize to steam directly downstream from the principal flashing location, which results in a dry annulus and not material wastage (Ref. 35).

The other extreme, the classic boric acid corrosion model studied in the set of Combustion Engineering experiments that were mentioned in Section 3.1, involves borated water dripping onto a hot metal surface. The accumulated solution is concentrated as the water boils off, and enhanced by oxygen available from the ambient atmosphere. The evaporation process locally cools the metal surface to the boiling point of concentrated boric acid solution, which turns out to be the temperature at which boric acid corrosion is at a maximum. The extent of cooling is a function of the leak rate, and in the case of Davis-Besse, the leak rate from VHP Nozzle #3 was sufficiently high to cool the head and allow for the boric acid solution to cover the walls of the cavity. From an inspection of these extremes, it is apparent that the rate of leakage from VHP Nozzle #3 would have had to have been high (> 0.1 gpm) for the observed wastage to occur (Ref. 35).

The Davis Besse root cause report offers a scenario that endeavors to explain the progression of events based on the available data. However, the differences between the wastage at Davis-Besse and incidences of VHP cracking at other plants are still unclear.

As mentioned previously, many scenarios were developed that describe possible mechanisms and conditions that could have led to an increased rate of RPV head wastage. Yet, these scenarios were largely unsubstantiated by experimental data. In response to this need to expand the state of knowledge with respect to corrosion and wastage of RPV head steels in concentrated boric acid solutions, the NRC and ANL conducted electrochemical potential and corrosion rate experiments (Ref. 35).

The NRC/ANL program produced the needed experimental data for electrochemical potentials (ECPs) and corrosion rates of the A533 Grade B low-alloy steel vessel head, the Alloy 600 VHP nozzle, and the 308 stainless steel (used in reactor pressure vessel cladding) weld clad in varying concentrations of boric acid solutions at temperatures between 95–316 °C (203–600 °F). ANL performed ECP tests on A533 Grade B steel, Alloy 600, and 308 SS to determine the effect of galvanic corrosion on the wastage of these materials. The galvanic differences between these materials do not point to galvanic corrosion as a strong contributor to the overall wastage process.

The corrosion rate tests simulated a range of possible annulus conditions, including (1) low leakage rates through the nozzle with the annulus plugged, (2) low leakage rates through the nozzle with the annulus open, and (3) high leakage rates that result in substantial local cooling. These annulus conditions were simulated using a variety of chemical environments, including (1) high-temperature, high-pressure aqueous boric acid solutions of varying concentrations, (2) high-temperature molten salt solutions, and (3) low-temperature (approximately 95 °C) boric acid solutions of varying concentrations including saturated. Only the low-alloy steel exhibited any appreciable wastage. The Alloy 600 and 308 stainless steel (used in reactor pressure vessel cladding) experienced virtually no wastage (Ref. 35). The joint NRC/ANL program produced valuable experimental data that may aid in the diagnosis and prevention of future incidences of boric acid corrosion. One of the most important findings of this program is that the corrosion rate of A533 Grade B steel in humidified molten salts of H-B-O (a mixture of

30

boric acid species and water) can cause wastage at rates as high as those for saturated aqueous solutions.

The industry Materials Reliability Program (MRP), headed by EPRI, was designed with objectives similar to the NRC boric acid program and is scheduled for completion in 2006. The objectives of the industry program are to further the understanding of the progression of boric acid wastage at RPV head penetrations, determine plant specific parameters that may influence wastage, and support the development of required inspection intervals for PWR plants of varying head designs. The program encompasses four tasks: (1) a heated crevice test to determine which species in the H-B-O system is present during stagnant and low-flow conditions and to determine the influence of these species on corrosion rates, (2) a flowing loop test to analyze moderate- and high-flow conditions, which can perform real time corrosion rate and electrochemical potential tests under laminar and impact flow, (3) a series of separate effects tests to determine the corrosion rates of conditions that have not been studied to a great extent such as galvanic coupling and contact with molten boric acid, and (4) a series of full scale mock-up tests to determine corrosion rates under representative VHP nozzle leakage conditions including leak rates ranging from 0.0001 to 0.3 gpm, controlled thermal conditions, full size nozzles, simulated crack geometries, and interference fits (Ref. 58).

The results of both the industry and NRC/ANL boric acid test programs will be included in Revision 2 of the EPRI Boric Acid Guidebook (BAGB) scheduled for publication in 2007. The original BAGB, published in 1995 and revised in 2001, was designed to provide support to licensees in their evaluation of ongoing degradation. It is a compilation of the utilities' boric acid corrosion experience and the boric acid corrosion programs that were completed at the time the guidebook was published (Ref. 59).

4.0 SUMMARY

This report includes a summary of foreign and domestic Alloy 600 cracking experience, an analysis of the Alloy 600 cracking susceptibility model for vessel head penetration (VHP) nozzles, and a collection of information on corrosion of pressure boundary materials in boric acid solutions.

The impetus for collecting information on Alloy 600 cracking was the March 2002 discovery of the VHP flaws, leaks, and pressure boundary corrosion at the Davis-Besse plant. The survey of Alloy 600 suggests that Alloy 600 and its associated welds (Alloy 182 and Alloy 82) are susceptible to crack nucleation and growth in a wide range of applications. Cracking has not been limited to VHPs. In the United States, significant findings in components other than VHPs have been observed at V.C. Summer (hot leg "A" cracks); South Texas Project 1 (BMI nozzle cracks); North Anna 2 (well-developed, outside diameter, circumferential cracks in a penetration not exhibiting leakage products), and in pressurizer heater sleeves at several plants.. Alloy 600 will continue to crack during the service life of a component. The intent of inspection is to identify and remediate the cracking before it can challenge safety systems.

The NRC issued Order EA-03-009 as part of the response to the Davis-Besse event. The Order established interim inspection requirements for reactor pressure vessel (RPV) heads depending upon their EDYs. The most extensive use of the EDY formula is in the prediction of the occurrence of stress-corrosion cracking (SCC) in Alloy 600 components including VHP nozzles. Within the limits specified by the Order, the EDY susceptibility model is only applicable to the upper reactor vessel head (and not the lower head) because the lower head is at a cooler temperature. Without these limits, however, the EDY model is applicable to all Alloy 600 components. The initial EDY formula, or susceptibility model, incorporated time, temperature, and stress. Other parameters such as hardness, yield strength, and carbide coverage were considered, but ultimately discarded. As a result of a lack of laboratory data and difficulties in computation, only time and temperature were retained in the model. Since the initial presentation of the EDY formula, an increased amount of laboratory data and plant experience suggest the possibility of improvement. The current EDY model was used because there was an urgency to develop a tool to measure a plant's EDY after the Davis-Besse event. This model was a logical choice since it reasonably reflected industry cracking data, while taking a conservative approach.

VHP nozzle inspection results indicate that the EDY model is effective in designating groups of plants for specific types of inspections, and do not indicate a need for revising the EDY model in the short term. The EDY calculation is becoming less of an issue for high susceptibility plants as a result of reactor vessel head replacement, but is still important for moderate susceptibility plants. Industry experience suggests that the cost of 100 percent volumetric inspection of heads is greater than the cost of head replacement both financially and in terms of worker dose. Plants that are replacing their heads with heads featuring Alloy 690 penetrations will fall into the low-EDY susceptibility category initially and then accumulate EDY as the plant operates with the new head.

Because the Alloy 600 cracking issue has not been fully characterized, there are many foreign and domestic research projects in progress. Test programs for thick section Alloy 690 and its companion weld metals, Alloy 152/52 are scarce, but several are scheduled to begin in 2005. Additional work on the crack growth rates of Alloys 600 and 690 and their associated weld materials is underway in the Westinghouse Science Technology Center. As for the future of Alloy 600 in PWRs, it is likely that parts, components, and joints fabricated from Alloy 600 and weld filler metals Alloy 82 and 182 will continue to crack during operation. Implementing effective inspection programs will be crucial to identify and remediate cracking.

Boric acid leakage is a consequence of Alloy 600 cracking. This leakage can lead to boric acid corrosion of the low-alloy steel that is in contact with the aqueous solution or dried solution products. Boric acid corrosion poses a considerable maintenance problem for licensees of PWRs and, in some cases, has impacted the integrity of low-alloy and carbon steel components. In response to the continued occurrence of boric acid wastage, which has raised concern within the agency and industry, the NRC has issued several GLs, bulletins, and INs, as well as an Order to address this issue. The occurrence of RPV head degradation by boric acid corrosion has changed the emphasis of many inspection plans from leakage-based inspection to crack detection before leakage.

There is more than 35 years of industry experience with this degradation mechanism, however, there is still a large deficit of knowledge with respect to the conditions and mechanisms that produce this phenomenon. The Davis-Besse RPV head wastage that occurred in the years leading up to March 2002 has spurred renewed interest into research of boric acid corrosion and has shown that this process is still not well understood. Both the NRC and industry have implemented boric acid corrosion test programs.

The Office of Nuclear Regulatory Research (RES) has initiated several programs. Pieces of the Davis-Besse reactor vessel head have been removed and shipped to Pacific Northwest National Laboratory (PNNL), where they will be examined. At Argonne National Laboratory (ANL), an extensive program of corrosion testing of several boundary materials in concentrated solutions of boric acid has been completed. The results of the NRC boric acid corrosion test program have shown that the galvanic difference between A533 Grade B steel, Alloy 600, and 308 stainless steel (used in reactor pressure vessel cladding) is not significant enough to consider galvanic corrosion as a strong contributor to the overall boric acid corrosion process. In addition, the NRC test program has revealed that the corrosion rate of A533 Grade B steel in humidified molten salts of H-B-O (a mixture of boric acid species and water) can cause wastage at rates as high as those for saturated aqueous solutions.

The industry program, headed by the Electric Power Research Institute (EPRI), was designed with similar objectives as the NRC boric acid program. The industry results are expected to be completed in 2006. The results of both the industry and NRC/ANL boric acid test programs will be included in Revision 2 of the EPRI Boric Acid Guidebook (BAGB) scheduled for publication in 2007.

5.0 REFERENCES

1.   BWXT Services Report 1140-025-02-24, "Final Report: Examination of the Reactor Vessel (RV) Head Degradation at Davis-Besse," June 2003, available in ADAMS under Accession #ML032310058 and #ML032310060.

2.   H. Coriou, et al., "Sensitivity to Stress Corrosion and Intergranular Attack of High-Nickel Austenitic Alloys," *Corrosion* (22), pp. 280–290 (1966).

3.   H.R. Copson and S.W. Dean, "Effect of Contaminant on Resistance to Stress-Corrosion Cracking of Ni-Cr Alloy 600 in Pressurized Water," *Corrosion* (21), pp. 1–8, 1966.

4.   J. Blanchet et al. (with H. Coriou), "Historical Review of the Principal Research Concerning the Phenomena of Cracking of Nickel-Base Austenitic Alloys," *Proceedings of the Conference in Fundamental Aspects of Stress-Corrosion Cracking and Hydrogen Embrittlement of Iron-Base Alloys, Unieux-Firminy, France, June 12–16, 1973*, National Association of Corrosion Engineers, Houston, TX.

5.   "Crack Growth Rates for Evaluating Primary Water Stress-Corrosion Cracking (PWSCC) of Thick-Wall Alloy 600 Materials (MRP-55)," Revision 1, EPRI-TR-1006695, prepared by G. White, Dominion Engineering, Inc., 2002.

6.   "Literature Survey of Cracking of Alloy 600 Penetrations in PWRs," EPRI NP-7094, prepared by A.S. O'Neill and J.F. Hall, Combustion Engineering, Windsor, CT, 1990.

7.   *Proceedings: 1991 EPRI Workshop on PWSCC of Alloy 600 in PWRs, October 9–11, 1991, Charlotte, NC*, EPRI TR-100852, prepared by Dominion Engineering, Inc., McLean, VA 22101.

8.   *Proceedings: EPRI Workshop on PWSCC of Alloy 600 in PWRs, December 1–3, 1992, Orlando, FL*, EPRI TR-103345, prepared by Dominion Engineering, Inc., McLean, VA.

9.   "PWSCC of Alloy 600 Materials in PWR Primary System Penetrations," EPRI TR-103696, Prepared by E.S. Hunt and D.J. Gross, Dominion Engineering, Inc., McLean, VA, 1994.

10. *Proceedings: EPRI Workshop on PWSCC of Alloy 600 in PWRs, November 15–17, 1994, Tampa, FL*, EPRI TR-105406, prepared by Dominion Engineering, Inc., McLean, VA.

11. *Proceedings of the IAEA Specialists' Meeting on Cracking in LWR RPV Head Penetrations, May 2–3, 1995, Philadelphia, PA*, NUREG/CP-0151, prepared by C.E. Pugh, Oak Ridge National Laboratory, Oak Ridge, TN.

12. *Proceedings: EPRI Workshop on PWSCC of Alloy 600 in PWRs, February 25-27, 1997, Daytona Beach, FL*, EPRI TR-109138, prepared by Dominion Engineering, Inc., McLean, VA.

13. *Proceedings: EPRI Workshop on PWSCC of Alloy 600 in PWRs (PWRMRP-27), February 14–16, 2000, At. Petersburg Beach, FL*, EPRI TR-1000873, prepared by Dominion Engineering, Inc., McLean, VA.

14. "Crack Growth of Alloy 182 Weld Metal in PWR Environments (MRP-21)," prepared by W. Bamford and J. Foster, Westinghouse Electric Corp., Pittsburgh, PA, June 2000.

15. C. Jansson, "Ringhals 4 Safe End Cracking in Alloy 182," presentation at ICG-EAC in Kyongju, Korea, April 23–27, 2001.

16. *Proceedings: EPRI Workshop on PWSCC of Alloy 600 in PWRs, June 13–14, 2001, Atlanta, GA*, EPRI TR-1006278, prepared by F. Ammirato, EPRI NDE Center, Charlotte, NC.

17. NISA/METI Press Release, "Leakage from Piping Nozzle Stub for Installment of Control Rod Drive Mechanism, etc. Found During the Periodical Inspection of Ohi Unit 3," May 6, 2004.

18. W.T. Russell, Associate, letter to W. Rasin, "Safety Evaluation for Potential Reactor Vessel Head Adaptor Tube Cracking," 1993.

19. BAW-10190, Addendum 1, "External Circumferential Crack Growth Analysis for B&W-Design Reactor Vessel Head Control Rod Drive Mechanism Nozzles," December 1993.

20. Topical Report CEN-614, "Safety Evaluation of the Potential for and Consequence of Reactor Vessel Head Penetration Alloy 600 OD-Initiated Nozzle Cracking," December 1993.

21. Generic Letter 97-01, "Degradation of Control Rod Drive Mechanism Nozzle and Other Vessel Closure Head Penetrations," U.S. Nuclear Regulatory Commission, Washington, DC, 1997.

22. D.J. Modeen (Nuclear Energy Institute), letter to G.C. Lainas (U.S. Nuclear Regulatory Commission), "Responses to NRC Requests for Additional Information on Generic Letter 97-01," December, 11, 1998.

23. Information Notice 2001-05, "Through-Wall Circumferential Cracking of Reactor Pressure Vessel Head Control Rod Drive Mechanism Penetration Nozzles at Oconee Nuclear Station, Unit 3," U.S. Nuclear Regulatory Commission, Washington, DC, April 30, 2001.

24. Bulletin 2001-01, "Circumferential Cracking of Reactor Pressure Vessel Head Penetration Nozzles," U.S. Nuclear Regulatory Commission, Washington, DC, August 3, 2001.

25. Bulletin 2002-02, "Reactor Pressure Vessel Head and Vessel Head Penetration Nozzle Inspection Programs," U.S. Nuclear Regulatory Commission, Washington, DC, 2002.

26. Order EA-03-009, "Issuance of Order Establishing Interim Inspection Requirements for Reactor Pressure Vessel Heads at Pressurized-Water Reactors," U.S. Nuclear Regulatory Commission, Washington, DC, February 11, 2003.

27.   Order EA-03-009, Revision 1: "Issuance of First Revised NRC Order (EA-03-009) Establishing Interim Inspection Requirements for Reactor Pressure Vessel Heads at Pressurized-Water Reactors," U.S. Nuclear Regulatory Commission, Washington, DC, February 20, 2004.

28.   Information Notice 2003-11, "Leakage Found on Bottom-Mounted Instrumentation Nozzles," U.S. Nuclear Regulatory Commission, Washington, DC, 2003.

29.   Information Notice 2003-11, Supplement 1, "Leakage Found on Bottom-Mounted Instrumentation Nozzles," U.S. Nuclear Regulatory Commission, Washington, DC, 2004.

30.   Bulletin 2003-02, "Leakage from Reactor Pressure Vessel Lower Head Penetrations and Reactor Coolant Pressure Boundary Integrity," U.S. Nuclear Regulatory Commission, Washington, DC, 2003

31.   T. Jossang, "Energetics in Dislocation Motion," *Energetics in Metallurgical Phenomena, Proceedings of the 1962 Seminar*, W.M. Mueller, ed., Gordon and Breach.

32.   "PWR Materials Reliability Program Response to NRC Bulletin 2001-01 (MRP-48)," EPRI TR-1006284, Final Report, A.R. McIlree, Project Manager, August 2001.

33.   "Davis-Besse Confirmatory Action Letter Response: Root Cause Analysis Report," ADAMS Accession #ML021130029, April 19, 2002.

34.   *Boric Acid Corrosion Guidebook, Revision 1: Managing Boric Acid Corrosion Issues at PWR Power Stations*, Electric Power Research Institute, Palo Alto, CA, 2001.

35.   J.-H. Park, O.K. Chopra, K. Natesan, and W.J. Shack, "Boric Acid Corrosion of Light-Water Reactor Pressure Vessel Head Materials," NUREG/CR to be published in 2005, U.S. Nuclear Regulatory Commission, Washington, DC, 2004.

36.   Information Notice 80-27, "Degradation of Reactor Coolant Pump Studs," U.S. Nuclear Regulatory Commission, Washington, DC, June 11, 1980.

37.   Information Notice 82-06, "Failure of Steam Generator Primary Side Manway Closure Studs," U.S. Nuclear Regulatory Commission, Washington, DC, March 12, 1982.

38.   IE Bulletin 82-02, "Degradation of Threaded Fasteners in the Reactor Coolant Pressure Boundary of PWR Plants," U.S. Nuclear Regulatory Commission, Washington, DC, June 2, 1982.

39.   Information Notice 86-108, "Degradation of Reactor Coolant System Pressure Boundary Resulting from Boric Acid Corrosion," U.S. Nuclear Regulatory Commission, Washington, DC, December 29, 1986.

40.   Information Notice 86-108, Supplement 1, "Degradation of Reactor Coolant System Pressure Boundary Resulting from Boric Acid Corrosion," U.S. Nuclear Regulatory Commission, Washington, DC, April 20, 1987.

41. Information Notice 86-108, Supplement 2, "Degradation of Reactor Coolant System Pressure Boundary Resulting from Boric Acid Corrosion," U.S. Nuclear Regulatory Commission, Washington, DC, November 19, 1987.

42. Information Notice 86-108, Supplement 3, "Degradation of Reactor Coolant System Pressure Boundary Resulting from Boric Acid Corrosion," U.S. Nuclear Regulatory Commission, Washington, DC, January 5, 1987.

43. J.F. Hall, "A Survey of the Literature on Low-Alloy Steel Fastener Corrosion in PWR Power Plants," EPRI NP-3784, December 1984, also "Boric Acid Corrosion of Carbon and Low -Alloy Steel Pressure-Boundary Components in PWRs," August 1988.

44. Generic Letter 88-05, "Boric Acid Corrosion of Carbon Steel Reactor Pressure Boundary Components in PWR Plants," U.S. Nuclear Regulatory Commission, Washington, DC, March 17, 1988.

45. B&W Nuclear Service Company, "Safety Evaluation for B&W Design Reactor Vessel Head Control Rod Drive Mechanism Nozzle Cracking," B&W Owners Group Materials Committee, BAW-10190P, May 1993.

46. Bulletin 2002-01, "Reactor Pressure Vessel Head Degradation and Reactor Coolant Pressure Boundary Integrity," U.S. Nuclear Regulatory Commission, Washington, DC, March 18, 2002.

47. "Regulatory Issue Summary 2003-13: NRC Review of Responses to Bulletin 2002-01, 'Reactor Pressure Vessel Head Degradation and Reactor Coolant Pressure Boundary Integrity'," U.S. Nuclear Regulatory Commission, Washington, DC, July 29, 2003.

48. Information Notice 2003-02, "Recent Experience with Reactor Coolant System Leakage and Boric Acid Corrosion," U.S. Nuclear Regulatory Commission, Washington, DC, January 16, 2003.

49. W.H. Cullen and M.A. Switzer, "Degradation of Vessel Head Report," U.S. Nuclear Regulatory Commission, Washington, DC, ADAMS Accession #ML031110266, October 15, 2002.

50. C.J. Czajkowski, "Survey of Boric Acid Corrosion of Carbon Steel Components in Nuclear Plants," NUREG/CR-5576, U.S. Nuclear Regulatory Commission, Washington, DC, 1990.

51. R. Kilian, P. Scott, A. Roth, U. Wesseling, H. Venz, "Boric Acid Corrosion: European Experience," EPRI Boric Acid Corrosion Workshop, Baltimore, MD, July 25–26, 2002.

52. C.J. Czajkowski, "Boric Acid Corrosion of Ferritic Components," NUREG/CR-2827, U.S. Nuclear Regulatory Commission, Washington, DC, 1982.

53. "Boric Acid Corrosion of a Control Rod Drive Mechanism Flange Fastening Assembly Caused by a Deteriorated Gasket Results in Reactor Cool Coolant [sic] System Pressure Boundary Degradation," ADAMS Accession #ML9001120365, Event Date: December 8, 1989, LER Date: January 8, 1990.

54. Information Notice 1994-63, "Boric Acid Corrosion of Charging Pump Casing Caused by Cladding Cracks," U.S. Nuclear Regulatory Commission, Washington, DC, August 30, 1994.

55. "Excessive Corrosion of Incore Instrumentation Flange Components," ADAMS Accession #ML9408040213, Event Date: February 21, 1994.

56. "Sequoyah, Unit 2: Technical Assessment of Minor Reactor Vessel Head Material Wastage," ADAMS Accession #ML030070418, January 14, 2003.

57. Licensee Event Report (LER) No. 2003-002-00, "Reactor Coolant System Pressure Boundary Leakage Due To Degradation of an Alloy 600 Pressurizer Heater Bundle Diaphragm Plate," U.S. Nuclear Regulatory Commission, Washington, DC, ADAMS Accession #ML033580625, December 18, 2003.

58. "Materials Reliability Program Reactor Vessel Head Closure Penetration Safety Assessment for U.S. PWR Plants (MRP-110NP): Evaluation Supporting the MPR Inspection Plan," 1009807-NP, EPRI, Palo Alto, CA, 2004, ADAMS Accession #ML041680506, April 30, 2004.

59. C. Harrington, A. Ahluwalia, A. McIlree, J. Hickling, G. White, "Status of EPRI/MRP Boric Acid Corrosion (BAC) Testing Program," Meeting on Status of Research Activities, NRC Offices, Rockville, MD, March 22, 2004, ADAMS Accession #ML040500466, February 11, 2004.

APPENDIX A.

RECENT CRACKING AND LEAKAGE EVENTS

APPENDIX A.
RECENT CRACKING AND LEAKAGE EVENTS

The NRC Vessel Head Penetration (VHP) Conference included presentations that described several recent Alloy 600 cracking events. Tables 1a through 1d contain summaries of findings, gleaned from sources such as licensee event reports (LERs) and event notification reports submitted by U.S. licensees (B&W=Babcock & Wilcox, CE=Combustion Engineering, and W=Westinghouse). In this regard, only publicly available information is presented in this table. The findings are listed chronologically by the component that failed. This should provide an overall understanding of how Alloy 600 cracking events evolved over time, and the references are intended to provide traceability. A more detailed account of the individual incidents follows the table.

Table 1 Pressurized-Water Reactor Plants
Table 1a. Reactor Vessel Head Nozzles

Plant Type	Plant	Date	Component	Reference
B&W	ONS-1	November 2000	VHP Weld and thermocouple nozzles	LER 269-2000-006 Event Notification Report 37567
	ONS-3	February 2001	VHP and VHP Weld	LER 287-2001-001 Event Notification Report 37760
	ANO-1	March 2001	VHP	LER 313-2001-002 Event Notification Report 37864
	ONS-2	April 2001	VHP and VHP Weld	LER 270-2001-002 Event Notification Report 37950
	Crystal River 3	October 2001	VHP	LER 302-2001-004 Event Notification Report 38365
	TMI-1	October 2001	VHP	LER 289-2001-002 Event Notification Report 38416
	ONS-3	December 2001	VHP and VHP Weld	LER 287-2001-003 Event Notification Report 38493
	Davis-Besse	February 2002	VHP	LER 346-2002-002 Event Notification Report 38732
	ONS-1	April 2002	VHP and VHP Weld	LER 269-2002-003 Event Notification Report 38821
	ANO-1	October 2002	VHP and VHP Weld	LER 313-2002-003 Event Notification Report 39254
	ONS-2	October 2002	VHP	LER 270-2002-002 Event Notification Report 39288
	ONS-3	April 2003	VHP	LER 287-2003-001 Event Notification Report 39821
	ONS-1	September 2003	VHP and thermocouple penetration	LER 269-2003-002 Event Notification Report 40192
W 4-Loop	D.C. Cook 2	1994	VHP	30-day outage response to Bulletin 2001-01
	D.C. Cook 2	May 2003	Vessel head penetration	Event Notification Report 39855

Table 1a. Reactor Vessel Head Nozzles (continued)

Plant Type	Plant	Date	Component	Reference
W 3-Loop	Surry 1	Oct. Nov. 2002	VHP Weld	LER 280-2001-003 Event Notification Report 38435 30-day outage response to Bulletin 2001-01
	North Anna 2	November 2001	VHP Weld	LER 339-2001-003 Event Notification Report 38498
	North Anna 2	September 2002	VHP and VHP Weld	LER 339-2002-001 Event Notification Report 39191
	North Anna 1	February 2003	VHP	LER 338-2003-001 Event Notification Report 39635
W 2- Loop	///////	///////	///////	///////
CE	Millstone 2	March 2002	CEDM	Response to Bulletin 2002-02
	St. Lucie 2	April 2003	CEDM and CEDM weld	LER 389-2003-002 Event Notification Report 39812
CE Standard	///////	///////	///////	///////

/////// Indicates there have been no reports for this type of plant

Table 1b. Reactor Vessel Bottom Mounted Instrument Nozzles

Plant Type	Plant	Date	Component	Reference
B&W	///////	///////	///////	///////
W 4-Loop	STP-1	April 2003	Bottom mounted instrumentation	LER 498-2003-003 Event Notification Report 39754 ML032120244 ML032120667
W 3-Loop	///////	///////	///////	///////
W 2-Loop	///////	///////	///////	///////
CE	///////	///////	///////	///////
CE Standard	///////	///////	///////	///////

Table 1c. Pressurizer Nozzles

Plant Type	Plant	Date	Component	Reference
B&W	ANO-1	December 1990	Pressurizer instrumentation nozzle	LER 313-1990-021
	Crystal River 3	October 2003	Pressurizer level nozzle	LER 302-2003-003 Event Notification Report 40222
	TMI-1	November 2003	Pressurizer heater sleeve	LER 289-2003-003 Event Notification Report 40296
W 4-Loop	///////	///////	///////	///////
W 3-Loop	///////	///////	///////	///////
W 2-Loop	///////	///////	///////	///////

Table 1c. Pressurizer Nozzles (continued)

Plant Type	Plant	Date	Component	Reference
CE	San Onofre 3	February 1986	Pressurizer instrument nozzle	LER 362-1986-003
	St. Lucie 2	1987	Pressurizer instrument nozzles	Noted in LER 389-1993-004
	ANO-2	April 1987	Pressurizer heater sleeve	LER 368-1987-003
	Calvert Cliffs 2	May 1989	Pressurizer heater sleeves and level nozzle	LER 318-1989-007
	San Onofre 3	February 1992	Pressurizer vapor space level instrument nozzles	LER 361-1992-004
	San Onofre 2	March 1992	Pressurizer vapor space level instrument nozzles	LER 361-1992-004
	St. Lucie 2	March 1993	Pressurizer instrument nozzles	LER 389-1993-004
	Palisades 1	October 1993	Pressurizer temperature nozzle	LER 255-1993-011
	Calvert Cliffs 1	February 1994	Pressurizer heater sleeve	LER 317-1994-003
	St. Lucie 2	March 1994	Pressurizer instrument nozzles	LER 389-1994-002
	San Onofre 3	July 1995	Pressurizer level instrumentation nozzle	LER 362-1995-001
	San Onofre 2	March 1997	Pressurizer temperature nozzle	LER 361-1997-004
	Calvert Cliffs 1	1998	Pressurizer heater Sleeve	MRP-27
	Calvert Cliffs 2	July 1998	Pressurizer level tap	LER 318-1998-005
	Waterford 3	February 1999	Pressurizer instrument nozzles	LER 382-1999-002 Event Notification Report 35407
	San Onofre 3	1999	Pressurizer heater sleeve	MRP-27
	ANO-2	July 2000	Pressurizer heater sleeve	LER 368-2000-001 Event Notification Report 37199
	Ft. Calhoun	October 2000	Pressurizer Temperature Nozzle	ML003781299 ML023610110
	Waterford 3	October 2000	Pressurizer heater sleeve	LER 382-2000-011 Event Notification Reports 37422 and 37434
	Millstone 2	February 2002	Pressurizer heater sleeves	LER 336-2002-001
	ANO-2	April 2002	Pressurizer heater sleeve	LER 368-2002-001 Event Notification Report 38855, 38888

Table 1c. Pressurizer Nozzles (continued)

Plant Type	Plant	Date	Component	Reference
	Millstone 2	October 2003	Pressurizer heater sleeves	LER 336-2003-004
	Waterford 3	October 2003	Pressurizer heater sleeve	Event Notification Report 40278
CE Standard	Palo Verde 1	January 1992	Pressurizer steam-space nozzle	LER 528-1992-001
	Palo Verde 2	October 2000	Pressurizer heater sleeve	LER 529-2000-004 Event Notification Report 37411
	Palo Verde 3	September 2001	Pressurizer heater sleeve	Event Notification Report 38332
	Palo Verde 3	March 2003	Pressurizer heater sleeve	LER 530-2003-002 Event Notification Report 39714
	Palo Verde 3	February 2004	Pressurize heater sleeve	Event Notification Report 40556
Foreign	Tsuruga 2	September 2003	Pressurizer piping nozzle stub weld	http://www2.jnes.go.jp/atom-db/en/index.html

Table 1d. Reactor Coolant System Nozzles and Other Components

Plant Type	Plant	Date	Component	Reference
B&W	ANO-1	February 2000	RCS hot leg instrumentation	LER 313-2000-003 Event Notification Report 36697
W 4-Loop	Catawba 2	September 2001	SG bowl drain	LER 414-2001-002
	Catawba 2	September 2004	SG bowl drain	Event Notification Report 41048
W 3-Loop	Summer 1	October 2000	Hot leg nozzle	LER 395-2000-008 Event Notification Report 37423 http://www.nrc.gov/reactors/operating/ops-experience/Alloy600/vcsummer.html
W 2-Loop	////////			
CE	Palisades 1	September 1993	Relief valve	LER 255-1993-009
	San Onofre 3	July 1995	RCS hot leg instrument nozzles	LER 362-1995-001
	St. Lucie 2	October 1995	RCS hot leg instrument nozzle	LER 389-1995-004
	San Onofre 3	April 1997	RCS instrument nozzle	LER 362-1997-001
	San Onofre 3	July 1997	RCS instrument nozzle	LER 362-1997-002
	San Onofre 2	January 1998	RCS nozzles	LER 361-1998-002
	Waterford 3	February 1999	RCS hot leg instrument nozzle	LER 382-1999-002 Event Notification Report 35407

Table 1d. Reactor Coolant System Nozzles and Other Components (continued)

Plant Type	Plant	Date	Component	Reference
	ANO-2	July 2000	RCS hot leg instrument nozzle	LER 368-2000-001 Event Notification Report
	Waterford 3	October 2000	RCS hot leg mechanical nozzle seal assembly clamps	LER 382-2000-011 Event Notification Reports 37422 and 37434
	St. Lucie 1	April 2001	RCS hot leg instrument nozzle	LER 335-2001-003 Event Notification Report 37919
	Waterford 3	October 2003	RCS hot leg instrument nozzle	Event Notification Report 40277
CE Standard	Palo Verde 1	October 1999	RCS hot leg valves	LER 528-1999-006 Event Notification Report 36256
	Palo Verde 1	March 2001	RCS hot leg thermowell	LER 528-2001-001 Event Notification Report 37878
	Palo Verde 3	September 2001	RCS hot leg temperature nozzle	Event Notification Report 38332
	Palo Verde 3	March 2003	RCS hot leg instrument nozzle	LER 530-2003-002 Event Notification Report 39714

Table 2. Boiling-Water Reactor Plants — All Components

Plant Type	Plant	Date	Component	Reference
GE Type 4	Vermont Yankee	April 1986	Core spray nozzle weld	LER 271-1986-005
	Hope Creek 1	September 1997	Core spray nozzle weld	LER 354-1997-023
	Duane Arnold 1	November 1999	Recirculation riser welds	LER 331-1999-006 Event Notification Report 36416 and 36402
	Susquehanna 1	March 2004	Recirculation weld	Event Notification Report 40605
GE Type 3	Pilgrim 1	October 2003	Reactor vessel nozzle to cap weld	LER 293-2003-006

Reactor Vessel Head Nozzles

B&W Plants

Oconee Unit 1 (Event Notification Report 37567 + LER 269-2000-006)

On November 25, 2000, during a visual inspection of the top surface of the reactor pressure vessel head, small amounts of boric acid deposited on the vessel head surface was discovered. These deposits appeared to be located at the base of five unused thermocouples and the Nozzle #21 weld at points where they all penetrate the RPV head surface. On December 4, an eddy current test was performed on the inside surface of the eight thermocouple nozzles and revealed axial crack-like indications on the ID of the nozzles in the vicinity of the partial penetration weld. Dye penetrant testing on Nozzle #21 identified two very small pin-hole indications running at a slightly skewed angle across the fillet weld. The eight thermocouple nozzles were removed and had Alloy 690 plugs welded in place. The indications in the VHP fillet weld were ground out, and a final weld repair was performed.

Oconee Unit 3 (Event Notification Report 37760 + LER 287-2001-001)

On February 18, 2001, during a visual inspection of the reactor vessel VHP nozzles, small amounts of boron residue surrounding the base of several control rod drive mechanism head penetrations was discovered. The boric acid deposits were identified around nozzles 3, 7, 11, 23, 28, 34, 50, 56, and 63. Subsequent surface dye penetrant test inspections of the weld areas and outside diameter identified several deep axial cracks that initiated near the toe of the fillet weld and propagated radially into the nozzle materials as well as axially along the outer diameter surface. Ultrasonic testing confirmed the existence of deep cracks in all nine leaking VHP nozzles. Of these 47 original crack indications, 19 were OD-initiated flaws that were not through-wall. There were 16 flaws that were OD-initiated through-wall cracks. There were nine circumferential flaws, with one being ID-initiated and the rest being OD-initiated. Two of the outer diameter circumferential flaws were above the J-groove weld. Finally, there were also three ID-initiated non-through-wall cracks.

Arkansas Nuclear Unit 1 (Event Notification Report 37864 + LER 313-2001-002)

On March 18, 2001, during a routine visual inspection of the reactor vessel head area, boric acid crystals were discovered. On March 24, 2001, eddy current testing and ultrasonic testing revealed a reactor coolant solution pressure boundary leak. The leak was identified in the wall of Nozzle #56. The UT data indicated that the crack was on the downhill side of the nozzle and extended approximately 0.8 inch below the weld and upward to approximately 1.0 inch above the weld. The depth of the crack was approximately 0.2 inch. The axial crack in the VHP was removed and an embedded flaw repair technique (basically a weld overlay that covers and seals off the pre-exisiting flaw) was utilized.

Oconee Unit 2 (Event Notification Report 37950+ LER 270-2001-002)

A preliminary reactor head visual inspection on April 28, 2001, revealed small amounts of boron residue surrounding nozzles 4, 16, 18, and 30. Subsequent surface dye penetrant test inspections of the weld area and nozzle outside diameter identified several axial cracks on four VHP nozzles that initiated near the toe of the fillet and propagated radially into the nozzle materials as well as axially along the outer diameter surface. Eddy current tests revealed two shallow axial flaws on Nozzle #16 and craze cracking on all four VHP nozzles' inner diameter surface. The ultrasonic testing confirmed the existence of some axial cracks with one short outer diameter initiated circumferential crack on VHP 18. The circumferential flaw was OD-initiated, extended 11 percent through-wall, and was 1.26 inch in length.

Crystal River Unit 3 (Event Notification Report 38365+ LER 302-2001-004)

On October 1, 2001, during a visual inspection of the reactor vessel VHP, boric acid buildup was discovered around Nozzle #32. Ultrasonic testing performed on Nozzle #32 revealed two through-wall axial cracks. These cracks extended from the bottom of the nozzle to above the J-groove weld. These two axial cracks then joined a circumferential crack above the J-groove weld. The circumferential flaw was about 90° and 50 percent through-wall. There was another circumferential flaw below the J-groove weld which extended 30° and was approximately 75 percent through-wall. Nozzle 32 was repaired using the ambient temperature bead repair technique.

Three Mile Island Unit 1 (Event Notification Report 38416 + LER 289-2001-002)

On October 11 and 12, 2001, during a visual inspection of the reactor vessel VHP nozzles, boric acid buildup was discovered around 8 different thermocouple nozzles. Liquid penetrant testing and ultrasonic testing identified through-wall indications on three VHP nozzles. These engineering evaluations concluded that the indications at Nozzles #44, 35, and 37 indicate a reactor coolant system pressure boundary leaks. During later examinations of the reactor vessel head, Nozzles #29 and 64 were shown to contain RCS pressure boundary leaks. The VHP nozzles were repaired by initially rolling the nozzle above the J-groove weld, and then machining the lower portion of the VHP nozzle including portions of the J-groove weld. A new pressure boundary weld was formed between the VHP nozzle and the RPV head low-alloy steel at the location above the previous J-groove weld and below the rolled nozzle area. A surface remediation inducing compressive stresses was performed after the repair. The thermocouples were repaired by cutting them approximately 1 inch from the outside surface of the RPV head. The remaining nozzle portion inside the RPV head was machined out of the head. These six thermocouple nozzles were plugged by installing an Alloy 690 plug in the RPV head bore. A nozzle weld dam was then inserted into the cavity. A weld pad buildup of Alloy 152 was welded over the nozzle plug.

Oconee Unit 3 (Event Notification Report 38493 + LER 287-2001-003)

On December 12, 2001, during a visual inspection of the reactor vessel VHP nozzles, leakage indications were discovered as evidenced by minor boric acid buildup around 4 VHP nozzles. Nondestructive examination of the suspected nozzles revealed that 7 of the 69 total nozzles required repair. Five of these seven nozzles had a leak pathway to the top of the reactor vessel head. Some of the indications were in the nozzles themselves, while other indications extended slightly into the weld. Most of the cracking was axial in nature, however, there was one circumferential flaw found in Nozzle #2 above the J-groove weld.

Davis-Besse Unit 1 (Event Notification Report 38732 updated March 6, 2002 and LER 346-200-2002)

On February 26, 2002, during a refueling outage, visual inspections were conducted. The inspections were inconclusive because of previous known boric acid deposits. On February 27, 2002, ultrasonic testing data identified axial through-weld indications on Nozzle #3. Engineering evaluations of this data confirmed reactor coolant system pressure boundary leakage exists. UT data for VHP Nozzles #1 and 2 exhibited axial indications that represented boundary leakage. Additionally, a circumferential indication on this VHP nozzle was 34••in circumferential length, and 50 percent through-wall. In the process of machining Nozzle #3, unexpected movement of the nozzle occurred. As a result of this movement and further investigation, a cavity in the RPV head was discovered. The width of this cavity measured approximately 4–5 inches at its largest point. The cavity formation is attributable to boric acid corrosion over a long period of time. The initial cracking of the VHP lead to leakage of reactor solution into the annulus. The combination of pressurized-water stress-corrosion cracking (PWSCC) and boric acid corrosion was the main contributor to this event.

Oconee Unit 1 (Event Notification Report 38821 + LER 269-200-2003)

On April 1, 2002, a qualified visual inspection of the Unit 1 reactor vessel head was conducted. Two penetrations had very slight amounts of boron accumulations. Ultrasonic testing was then performed on the penetrations and revealed partial through-wall outside diameter cracks for nozzles 1, 7, and 8. There were five flaws and a potential leak path identified in Nozzle #7. There was one axial flaw but no leak path identified in Nozzle #8. Nozzle #1 showed three minor indications in a region of rough weld contour. Liquid dye-penetrant was then used to examine these three nozzles. PT revealed two axial flaws in the original weld on Nozzle #7. PT was also used on Nozzle #1 and there were no recordable or rejectable PT indications. Nozzles #7 and 8 were repaired by removing part of the nozzles below the reactor vessel head (RVH) and a length of 5 inches into the RVH. A new pressure boundary weld was installed within the bore, inspected, and surface conditioned with a water-jet peening process.

Arkansas Nuclear One, Unit 1 (Event Notification Report 39254 + LER 313-2002-003)

On October 7, 2002, a routine visual inspection was performed on the reactor vessel head area. Small boric acid crystal nodules were found around the area of control rod drive mechanism Nozzle #56. On the downhill side of the nozzle, boric acid residue was located, extending 180° around the nozzle annulus area, with a small boric acid nodule at the most downhill point. Nondestructive examination (NDE) revealed indications of cracking in the nozzle, which was the cause of the boric acid residue. NDE of all the nozzles revealed indications of non-through-wall cracks in six other nozzles, and a likely porosity weld defect in another. The leaking Nozzle #56 had been repaired during the previous outage. The new crack indications were located just outside the previous weld repair zone. The previous repair technique was the embedded flaw repair. It is believed that the same nozzle failed because the previous repair did not isolate the 182 weld, which is a susceptible material to PWSCC in the pressurized-water reactor (PWR) environment. The current repair technique consisted of removing the portion of the nozzle that extends below the surface of the reactor vessel head. A new half-nozzle was installed using Alloy 52 weld material.

Oconee Unit 2 (Event Notification Report 39288 + LER 270-2002-002)

During a visual inspection on October 15, 2002, evidence of through-wall leakage was discovered on seven VHP penetrations. These penetrations were Nozzles #8, 9, 19, 24, 31, 42, and 67. None of these nozzles had been previously repaired. Additional nozzle head penetrations were masked by boric acid deposits suspected of being from separate sources of leakage. NDE was used to characterize the cracking. No circumferential cracks were reported and 10 VHPs (#11, 15, 19, 21, 24, 31, 33, 36, 38, and 42) with axial cracks were found. The repair technique included removing part of the previous nozzle and welding a new half-nozzle into the reactor head. The new nozzle was treated with water jet peening afterwards.

Oconee Unit 3 (Event Notification Report 39821 + LER 287-2003-001)

Unit 3 entered its scheduled end-of-cycle 20 refueling outage on April 20, 2003. During a visual inspection of the reactor vessel head on May 2, 2003, evidence of possible through-wall leakage was observed on two VHP penetrations. The locations of these penetrations are Nozzles #4 and 7. Nozzle #4 contained a very thin white coating while nozzle 7 appeared to have a small accumulation of boron on the head adjacent to the annulus region. Approximately 6 to 8 additional nozzles-to-head penetrations were masked by deposits from a component cooling system leak above the RV head and were unable to be inspected. Prior refueling outage RVH inspection videotapes showed that the Nozzle #7 deposits were not associated with a new leak but rather were remnants from a prior outage leak and repair where the boron residue had not been removed. The Nozzle #4 boron deposits were similar to previous RVH leaks. The apparent root cause of the nozzle leak is PWSCC. The head was replaced during this refueling outage.

Oconee Unit 1 (Event Notification Report 40192 + LER 269-2003-002)

During a scheduled bare metal visual inspection on September 23, 2003, possible evidence of a through-wall leak on two VHPs (Nozzles #6 and 16) and one thermocouple penetration (Nozzle #7) was observed. The thermocouple had been repaired (plugged) in December 2000. Reactor coolant leakage prior to the unit shutdown varied between 0.15 and 0.24 gallon per minute. There were no plans to perform additional inspections or repairs since the head was replaced during the same refueling outage.

Westinghouse 4-Loop Plants

Cook Unit 2 Alloy 600- Bulletin 2001-01 Plant Specific Information (30-day response)

During the cycle 10 refueling outage in 1994, eddy current testing examination was performed on 71 of the 78 vessel head penetrations. The testing showed crack indications in penetration number 75. Three indications were found with lengths of 9 mm, 16 mm, and 45 mm. These indications were axial in orientation and were closely spaced. The 3 indications were located near the 160-degree location on the high side. The 45-mm crack was located near the J-groove weld, but was mostly below the weld.

Cook Unit 2 [Event Notification Report 39855 updated on June 11, 2003 (retracted)]

Craze cracking indications were found on a reactor pressure vessel head penetration May 17, 2003. Shallow indications were found on the inside diameter of Nozzle #74 during the reactor head inspection. These indications are closely spaced • •inch below the J-groove weld. Initial calculations showed a crack depth of 0.117 inch. There was no through-wall leakage detected. These same cracking indications were found during the 2002 refueling cycle and have not shown any significant growth. This report was retracted because it was determined that the craze cracking indications in Nozzle 74 of the Unit 2 RPV head do not represent a seriously degraded principal safety barrier of the nuclear power plant.

Westinghouse 3-Loop Plants

Surry Unit 1 (Event Notification Report 38435 + LER 280-2001-003 + 30 day outage response to Bulletin 2001-01)

On October 28, 2001, through-wall indications of the J-groove weld were identified on VHP Nozzles #27 and 40. On November 2, 2001, indications of flaws in the penetration welds were also uncovered on Penetrations #65, 47, 69, and 18. The repair of these nozzles utilized the temper bead repair procedure.

North Anna Unit 2 (Event Notification Report 38498 + LER 339-2001-003)

On November 13, 2001, a through-wall leak on Nozzle #63 was observed. This event was treated as a through-wall leak based on the qualified visual inspection results and liquid penetrant examination. A portion of the weld around Nozzle #63 was excavated to a depth of approximately 1" of weld metal. The liquid penetrant exam of this excavation showed 12 indications located at the outside edge of the weld almost the full length of the excavation, which turned into the weld at the uphill and downhill ends of the excavation. Six of the recorded cracks were transverse to the weld, while the other six were parallel. Eddy current testing revealed a crack 31mm in length in the area of the attachment weld. Ultrasonic testing of the same crack revealed that it was less than 1 mm in depth and 14 mm in length. Nozzle #51 also had nearby boric acid residue on the reactor vessel head. A liquid penetrant test revealed 12 indications at the toe of the weld. Five of these indications were parallel while the rest were transverse. Eddy current testing of the nozzle's weld revealed six axial indications. These axial cracks were less than 2 mm in depth and ranged from 6 to 24 mm in length. Similarly, Nozzle #62 was also investigated. This nozzle had eight indications at the toe of the weld. Two of the cracks were parallel and six were transverse. Eddy current testing revealed two axial indications. The dimensions were 74 mm and 42 mm in length, while being less than 2 mm and less than 1 mm in depth, respectively. The repair used for the nozzles was the temper bead procedure.

North Anna Unit 2 (Event Notification Report #39191 + LER 339-2002-001)

Boric acid residue was discovered during a bare metal visual inspection of the reactor vessel head on September 14, 2002. It appears that Nozzles #21 and 31 had exhibited some leakage, as evidenced by boric acid residue on the reactor vessel head. Four additional penetrations were suspected of leaking, and several penetrations were masked with boric acid residue. Of the 59 J-groove weld penetrations that were inspected using eddy current testing (ET), 57 were identified with crack-like indications. The ET identified at least one indication of a 6-mm crack in about 83 percent of the J-groove welds. During the previous year's outage, no boric acid residue was discovered. The six nozzles (N2-51, 53, 55, 57, 62, and 63) that could not be inspected with ET had their welds inspected using liquid penetrant tests (PTs). Three of these penetrations (N2-51, 62, and 63) had been previously repaired with weld overlay of the J-groove. Each of the six penetrations that were inspected with PT had evidence of rejectable indications.

Eddy current examinations of the J-groove welds showed indications of axial and circumferential cracking with respect to the welding direction. The range in length was from 0.12 inch to 7.0 inches. Some longer flaws were recorded, but they actually comprise a series of small flaws with very short distances between. Eddy current testing of the inside diameter surface showed twenty of thirty-five penetration tubes had axial indications. These indications were believed to be less than 0.12 inch deep. Four nozzles (#21, 31, 51, and 63) showed evidence of a leak path in the shrink fit area between the vessel head and the tube. Nozzles #51 and 63 had been identified as leaking in the Fall of 2001. Repairs of these penetrations had been improperly applied because the weld overlay repair did not extend out far enough to cover the previous NDE indications. The six penetration welds inspected with PT had greater than $^1/_{16}$ -inch linear flaw indications. The licensee replaced the reactor vessel head instead of making multiple repairs.

North Anna Unit 1 (Event Notification Report 39635 + LER 338-2003-001)

On February 22, 2003, Unit 1 entered a scheduled refueling outage. During this outage, visual inspection was performed on the reactor vessel head. On March 4, 2003, an apparent reactor vessel head through-wall leak was noted on Nozzle #50. The inspection was a followup to a previous inspection in 2001. Boric acid residue was found approximately ½ inch in diameter on the lower side of the penetration-to-head transition. There were no signs of wastage on the reactor vessel head. The Unit 1 reactor head was replaced during the 2003 refueling outage.

CE Plants

Millstone 2 (Response to Bulletin 2002-02)

On March 13, 2002, Dominion Nuclear Connecticut, Inc. (DNC) informed the NRC that analysis of ultrasonic inspection results from the previous three weeks had determined that flaw indications were in three nozzles as shown in Table 3. The portions of the nozzles containing these flaws were bored out, and "half-nozzle" repairs were completed. In this method, the flawed nozzles were bored out from the underside (wetted side) of the head, to a location approximately half-way through the head, leaving the old nozzle with a single-V, weld-prepped surface. A new section of Alloy 690 nozzle tube, similarly prepped, was inserted and welded into place. This repair procedure moves the pressure boundary from the inside surface to a point about midway through the thickness of the head.

Table 3. Flaw indications from Millstone 2 plant in March 2002.

Nozzle Number	21	34	50
Number of Axial Indications	5	4	0
Number of Circumferential Indications	1	2	2

St. Lucie Unit 2 (Event Notification Report 39812 updated May 6, 2003, and LER 389-2003-002)

On April 30, 2003, during a refueling outage, a defect in Nozzle #72 was found. St. Lucie Unit 2 had approximately 14.0 effective degradation years at the start of the 2003 refueling outage, therefore, this plant has a high susceptibility in accordance with Order EA-03-009. Visual inspection of the reactor pressure vessel was clean, with no evidence of leakage from the 102 RPVH penetrations or wastage on the RPVH surface. The UT inspection identified an axial crack in CEDM Nozzle #72. The defect outer diameter connected and extended into the nozzle and into the J-groove weld between the nozzle and reactor vessel head. The defect was an axial flaw, 0.28 inch deep and 0.96 inch long on the downhill side of the penetration. On May 2, 2003, a second defect was identified in CEDM Nozzle #18. The defect is also outer diameter connected and described as axial. It extended into the nozzle and through the J-groove weld between the nozzle and reactor vessel head. This second defect measured 0.26 inch deep and 2.98 inches long. It was also located on the downhill side of the penetration. Neither flaw extended through the wall of the nozzle. Neither nozzle had any evidence of leakage from the annulus between the nozzle and the reactor pressure vessel head associated with the indications.

Westinghouse 4-Loop System

South Texas Project Unit 1 (Event Notification Report 39754 updated May 22, 2003, and LER 498-2003-003)

During a bare metal inspection performed on the vessel bottom head on April 12, 2003, two potential leaks were identified. The leaks appeared to be associated with bottom-mounted instrumentation (BMI) penetrations 1 and 46. The BMI penetrations are inspected every outage and no residue had been discovered during the previous outage on November 20, 2002. There was a small amount of residue surrounding the outer circumference of the BMI penetrations where the nozzles meet the bottom of the reactor vessel. There did not appear to be any wastage. Approximately, 150 mg and 3 mg of residue were collected from penetrations number 1 and 46 respectively. The initial indication of boron and lithium in these samples suggested that they were RCS residue. A lithium isotopic analysis was conducted and the results confirmed that the precipitate was indeed from the RCS. An approximation of age was developed by conducting a cesium isotope study on the samples. The results of this test suggested that the solution was roughly 3–5 years of age. This, in turn, suggested a small leak rate because of the time to push the leakage through the annulus. UT testing of 57 nozzles and visual inspection of all 58 BMI nozzles was completed on May 23, 2003. Nozzles 1 and 46 contained a total of 5 cracks. No cracks were identified in any of the other BMI penetrations. Nozzle 1 contained 3 cracks which were all axial. Only one of these cracks provided a leak path from either the outside of the nozzle above the J-groove weld or inside the nozzle to the annulus. Nozzle #46 contained 2 axial cracks. Only one of the cracks in Nozzle #46 provided a leak path from the outside of the nozzle above the J-groove weld to the annulus. None of the cracks in Nozzle #46 extend to the inside of the nozzle. This is supported by the eddy current tests which revealed that only Nozzle #1 had a crack on the inside wall.

Pressurizer Nozzles

B&W Plants

Arkansas Nuclear One, Unit 1 (LER 313-1990-021)

On December 22, 1990, a potential reactor coolant system leak in the area of a pressurizer upper-level instrumentation nozzle was identified. During a followup inspection, it was verified that a very small leak existed at the nozzle. Nondestructive testing was conducted which confirmed the existence of a small axial crack in the nozzle inner surface which breached the outside diameter of the nozzle at the toe of the nozzle-to-vessel weld. A temporary repair was completed which initially deposited a weld pad on the shell OD around the nozzle penetration. The next step was to prepare a partial penetration weld in the pad and a new nozzle was installed into the penetration from the shell OD. This left a small gap between the original nozzle and the new nozzle.

Crystal River Unit 3 (Event Notification Report 40222 + LER 302-2003-003)

On October 4, 2003, during a routine visual inspection of the upper-level instrument tap nozzles, very small reactor coolant leaks were found on Nozzles RC-1-LT1, RC-1-LT2, and RC-1-LT3. The leakage evidence for RC-1-LT1 and RC-1-LT3 consisted of stains and boric acid residue. The evidence on RC-1-LT2 consisted only of stains on the pressurizer carbon steel shell. There was no evidence of leakage on any of the similar pressurizer nozzles. The last unidentified leak rate completed prior to plant shutdown was 0.15 gpm. The three pressurizer upper level instrument tap nozzles were repaired using a half-nozzle technique. This technique replaces half of the Alloy 600 nozzle with Alloy 690 using the similar metal weld (Alloy 52/152). This technique also moves the pressure boundary from the internal weld to an external location.

Three Mile Island Unit 1 (Event Notification Report 40296 + LER 289-2003-003)

On November 4, 2003, an inspection of the pressurizer heater bundle identified a primary leak at the lower pressurizer heater bundle diaphragm plate. Boric acid residue was found between the diaphragm plate and the cover plate. Initially the leak was thought to be coming from a seal weld. Nondestructive examination (NDE) determined that the leak path was through the edge of the pressurizer heater bundle diaphragm plate, and that there were six indications. Four of these indications were surface flaws not associated with a through-wall crack. The heater bundle was initially repaired by depositing a seal weld over the areas of the pressurizer heater bundle diaphragm plate. A leak was revealed in later testing at normal operating pressure and temperature. Because of this leak, the lower pressurizer heater bundle assembly including the pressurizer heater bundle cover plate was replaced with a new heater bundle assembly. The new diaphragm plate was constructed out of type 304 austenitic stainless steel. The diaphragm plate was welded to the pressurizer using Alloy 52 weld materials.

CE Plants

San Onofre Unit 3 (LER 362-1986-003)

On February 27, 1986, a pressure boundary leak was observed in a ¾-inch diameter pressurizer level instrument nozzle. Dye penetrant testing was utilized and revealed a crack extending from the end of the nozzle inside the pressurizer, • •inch outward through the RCS pressure boundary.

St. Lucie Unit 2 (Cited in LER 389-1993-004)

In 1987, during the replacement of four pressurizer steam-space instrument nozzles, it was determined that two nozzles had cracks but there was no evidence of leakage.

Arkansas Nuclear Unit 2 (LER 368-1987-003)

On April 24, 1987, the licensee declared an unusual event and initiated a reactor shutdown as a result of a suspected reactor coolant system pressure boundary leak of approximately 60 drops per minute from the area of the pressurizer vessel lower head. It was determined that the leakage source was the heater sleeve for the X1 pressurizer heater. Because of this leakage, there was some corrosion damage to the carbon steel pressurizer shell. The dimensions of this damage in the carbon steel were ½ inch in diameter and ¾ inch deep. While trying to remove the heater element from the X1 heater sleeve it was discovered that the heater sheath had ruptured. Similarly, another heater, T4, was found to have the same sheath damage. The final determination was that the X1 heater sheath failed which resulted in the damage to the X1 heater sleeve. The damage to the internal sleeve led to the damage of the pressure cladding weld. The initial trigger event was attributable to fabrication residual stresses in these Watlow heaters. All heaters manufactured by Watlow, except for X1 and T4, were removed. There were six available spare heaters that were installed. The other 15 empty heater sleeves were fitted with dummy heater plugs welded to the sleeves. The heater sleeves X1 and T4 were cut off approximately • •inch below the internal welded area. The rest of the sleeves were drilled out. Plugs were then inserted into the X1 and T4 holes and welded to the outside of the vessel utilizing a temper bead welding repair process.

Calvert Cliffs Unit 1 and 2 (LER 318-1989-007)

On May 5, 1989, an inservice inspection of the Unit 2 pressurizer revealed evidence of reactor coolant leakage from 28 of the 120 pressurizer vessel heater penetrations and one upper level nozzle. No evidence of leakage was found on the Unit 1 pressurizer heater penetrations or pressure/level penetrations. Additional inspections using dye penetrant and eddy current tests of 28 Unit 2 and 12 Unit 1 heater sleeves were conducted. Three sleeves from Unit 2 were destructively examined. All cracks were axial and determined to have minimal safety-significance. Reaming and repair operations associated with fabrication of the Unit 2 pressurizer appear to have contributed to the cause. The Unit 2 pressurizer was repaired by replacing 119 heater sleeves with dual Alloy 690 heater sleeves. One heater sleeve was sleeved and plugged with Alloy 690. All four upper level nozzles were replaced with Alloy 690.

San Onofre Unit 3 (LER 361-1992-004)

On February 18, 1992, during refueling outage cycle 6, a dye penetrant examination of the pressurizer vapor space level instrument nozzles revealed the presence of a through-wall crack. The examination was attributable to boric acid crystals being found near the nozzle previously. The leaking nozzle was replaced with a new nozzle made of Alloy 690. Liquid penetration testing was conducted on the three remaining vapor space nozzles. Much smaller indications on two of the three nozzles were revealed in these tests. These three nozzles were also replaced with nozzles fabricated with Alloy 690. The water space instrument nozzles were also visually checked and no signs of leakage were apparent.

San Onofre Unit 2 (LER 361-1992-004)

On March 14, 1992, an inspection on Unit 2 pressurizer vapor space level instrument nozzles was conducted. Boric acid crystals were found at two of the nozzles. An interim repair of the Unit 2 nozzles with Alloy 690 was implemented prior to startup. Inspection of the remaining water and vapor space nozzles showed no signs of leakage.

St. Lucie Unit 2 (LER 389-1993-004)

On March 2, 1993, water was discovered dripping onto the floor in containment near the pressurizer. Visual inspection revealed that four upper instrument nozzles were leaking at the entry fitting to the pressurizer. Liquid penetrant and eddy current test revealed axial cracking in the four steam-space nozzles extending into the surrounding weld area. The leaking pressurizer steam-space nozzles were removed and replaced with nozzles made of Alloy 690.

Palisades Unit 1 (LER 255-1993-011)

On October 9, 1993, inspection of the pressurizer upper temperature nozzle penetration TE-0101 was found to be leaking. Subsequent inspection of the lower temperature nozzle penetration TE-0102 was also found to be leaking. The leaks were attributable to cracking in the Alloy 600 nozzle material. The two nozzles were repaired by installing a weld pad on the outside pressurizer shell.

Calvert Cliffs Unit 1 (LER 317-1994-003)

On February 16, 1994, boron deposits were located on the pressurizer heater sleeve B-3 after removing insulation. On February 23, 1994, after removing more insulation, boron deposits were also discovered on pressurizer heater sleeve FF-1. Boroscopic and eddy current tests revealed a circumferential bulge approximately 0.5 inch long and 0.019 inch high (diametrical) in the area of the boric acid leaks. The leakage area showed evidence of surface metal smearing and cold work. Sleeve FF-1 had been reworked because of the presence of a stuck reamer. Heater sleeve B-3 and FF-1 were plugged with an Alloy 690 plug welded to the outer diameter of the pressurizer lower head.

St. Lucie Unit 2 (LER 389-1994-002)

On March 16, 1994, boric acid was observed on the exterior of the pressurizer steam-space C instrument nozzle during an inspection. Dye penetrant was utilized and identified indications at the A, B, and C steam-space instrument nozzle welds. The D instrument nozzle weld was acceptable. The unacceptable cracks were in the "J" weld between the Alloy 690 nozzle (replaced in 1993) and the clad on the inside of the pressurizer.

San Onofre Unit 3 (LER 362-1995-001)

On July 22, 1995, during inspection of the Alloy 600 and 690 (see the March 1992 event for Alloy 690 installation) instrument nozzles, one pressurizer level instrumentation nozzle was found with a small amount of boric acid crystals and oxidation present. Dye penetrant testing indicated crack initiation in the heat-affected zone of the weld butter. The Alloy 690 pressurizer nozzle piece interior did not have indications of PWSCC. All the vapor space instrument nozzles were planned to be replaced with Alloy 690 using Alloy 52 weld filler metal.

San Onofre Unit 2 (LER 361-1997-004)

On March 3, 1997, steam was observed emanating from the pressurizer. It was concluded that the leak was caused by PWSCC of Alloy 600 type materials of the pressurizer liquid temperature thermowell nozzle. The crack was oriented parallel to the long axis of the nozzle. The nozzle was removed and replaced with Alloy 690.

Calvert Cliffs Unit 1 (MRP-27)

Unit 1 heater sleeves were nickel plated in 1994. One nickel plated heater sleeve (B-1) was found leaking during the 1998 refueling outage. Ultrasonic testing revealed a short axial indication.

Calvert Cliffs Unit 2 (LER 318-1998-005)

On July 25, 1998, a steam leak was discovered at an upper-level instrument nozzle on the pressurizer. A dye penetrant test of the nozzle proved that the Alloy 690 nozzle did not contain a leak pathway. Ultrasonic examination of the vessel shell was performed to look for defects in the shell material. No defects were found. Because no leak path was found in the nozzle or shell material, it was postulated that the crack was in the Alloy 600-type weld filler material of the nozzle. The leaking nozzle was cut and the outer portion of the nozzle was removed. A weld pad of Alloy 690 was installed around the penetration and a new nozzle manufactured of Alloy 690 was inserted. This nozzle was welded to the pad with Alloy 690 filler material.

Waterford Unit 3 (Event Notification Report 35407 + LER 382-1999-002)

On February 25, 1999, during a routine visual inspection, evidence of reactor coolant system leakage was found on two Alloy 600 instrument nozzles located on the top head of the pressurizer. The leakage was in the annulus area where the nozzle penetrates the pressurizer head. The nozzles are welded on the inner diameter of the pressurizer and joined to instrument valves RC-310 and RC-311. The two leaking nozzles located on the pressurizer were repaired using welded nozzle replacements.

San Onofre Unit 3 (EPRI MRP-27)

In 1999, a cracked heater sleeve was identified by eddy current testing. The heater had failed, swelled, and stuck within the sleeve. The flaw was approximately 40 percent through-wall on the inner diameter of the sleeve near the attachment weld.

Arkansas Nuclear One, Unit 2 (Event Notification Report 37199 + LER 368-2000-001)

Twelve pressurizer heater sleeves were found to be leaking. On July 30, 2000, boron residue was discovered on the reactor coolant system pressurizer heater power cables. The boron came from leaks in heaters B2 and D2. B2 is on backup heater bank 4, and D2 is on backup heater bank 6. After removing insulation from the pressurizer, the licensee discovered 10 additional pressurizer heater sleeves that had previous leakage. Eddy current testing on two of the heater sleeves indicated that there was a single, through-wall, axial crack in both sleeves below the J-groove weld. These cracks initiated from the inside surface of the sleeves. Ultrasonic testing showed no cracks in the shell base metal. The pressurizers were repaired with an ASME Code-qualified process.

Ft. Calhoun (ML003781229 + ML023610110)

On October 22, 2000, during a walkdown inspection, leakage was detected from the lower pressurizer liquid space temperature Nozzle #TE-108 was detected. A weld technique was used to repair the nozzle.

Waterford Unit 3 (Event Notification Report 37442, 37434 + LER 382-2000-011)

On October 17, 2000, during a bare metal inspection of the pressurizer heater sleeve number F-4, a small amount of boric acid residue was discovered. This pressurizer was repaired by plugging the penetration.

Millstone Unit 2 (LER 336-2002-001)

On February 19, 2002, pressurizer heater penetrations and pressurizer instrument nozzle penetration were examined with visual inspection. Two heater sleeves showed indications of minor leakage because of the boron precipitates discovered on the outside of the penetrations. The cause of this event is through-wall cracks in the two pressurizer heater sleeves. The leaking heater sleeves were repaired using mechanical nozzle seal assembly clamps.

Arkansas Nuclear One, Unit 2 (Event Notification Report 38855, 38888+ LER 368-2002-001)

Six heater sleeves were found to be leaking in the pressurizer. Five of the leaking heater sleeves were discovered on April 15, 2002, while the other was found on April 30, 2002. On April 15, boron deposits were discovered on five pressurizer heater sleeve penetrations. On April 30, boron residue was observed around the sixth pressurizer heater sleeve. Since similar events had occurred in the July 2000 outage, no NDE was conducted. The leaking pressurizer heater sleeves were repaired using the mechanical nozzle seal assembly (MNSA) technique.

Millstone Unit 2 (LER 336-2003-004)

During the October 2003 outage, two leaking pressurizer heater penetrations were identified. These two pressurizer heaters, along with the two degraded pressurizer heaters found in the previous outage, were planned to be removed during the current outage. Ultrasonic testing determined that the flaws were axial in nature. The leaking heater penetrations were repaired using the MNSA technique.

Waterford Unit 3 (Event Notification Report 40278)

During an inspection on October 26, 2003, evidence of leakage was detected on pressurizer heater sleeves C-1 and C-3. The leakage was later determined to be boric acid. The two leaking heater sleeves were repaired using the second generation MNSA-2 technique

CE Standard Plants

Palo Verde Unit 1 (LER 528-1992-001)

On January 2, 1992, a pressure boundary leak was discovered in the pressurizer steam-space nozzle. A pad weld was put in place in order to stop the reactor coolant leakage. PWSCC is believed to be the cause of the leakage.

Palo Verde Unit 2 (Event Notification Report 37411 + LER 529-2000-004)

On October 4, 2000, during an inservice inspection, reactor coolant system pressure boundary leakage was discovered. The leakage was discovered at pressurizer heater nozzle sleeve A06. The leakage was detected in the form of small deposit of boron accumulation on the sleeve. Eddy current testing indicated linear axial cracking. The degraded heater sleeve was repaired by initially cutting off the heater sleeve close to the pressurizer bottom head. The degraded sleeve was then counter-bored and a reinforcing pad and plug were welded to seal the sleeve location. The repairs were made using Alloy 690 material. Another pressurizer was also repaired during the same refueling outage. This heater had failed and swelled in 1991. There was also a linear, axial indication in this sleeve. The sleeve was repaired in the same fashion as discussed before.

Palo Verde Unit 3 (Event Notification Report 38332)

On September 20, 2001, evidence of reactor coolant leakage was discovered. The leak was from a pressurizer heater sleeve nozzle. The pressurizer heater sleeve leakage is located at pressurizer heater B17. The leakage was identified by the discovery of boron deposits accumulated around the circumference of the pressurizer. There was no evidence of leakage during the last refueling outage. The MNSA technique was used for repairing the nozzle.

Palo Verde Unit 3 (Event Notification Report 39714 + LER 530-2003-002)

On March 29, 2003, engineering personnel performing preplanned visual examinations of reactor coolant system piping discovered boric acid on the RCS hot leg instrument nozzle and a pressurizer heater sleeve. There was boric acid residue discovered on the backup pressurizer heater sleeve A01. Eddy current testing on the heater sleeve suggested that the cracking was axial in nature. The heater sleeve was repaired using the MNSA technique.

Palo Verde Unit 3 (Event Notification Report 40556)

On February 29, 2004, engineering personnel were performing a visual examination of the RCS piping and discovered boric acid residue on the A03 pressurizer heater sleeve. The visual observation was characterized as a small white buildup of boron residue around the heater sleeve as the sleeve enters the pressurizer bottom head. There was no residue running down the outside of the sleeve, and there were no signs of dripping, spraying, puddles of liquid or liquid running down the nozzle or pressurizer. The residue appeared to be dry. The heater was repaired using the MNSA technique.

Foreign Plants

The following summary of the Alloy 600 cracking incident at Tsuruga, in Japan, is included because it closely resembles many cracking incidents in U.S. plants.

Tsuruga Unit 2

Cracking was discovered September 5, 2003, on the pressurizer relief piping nozzle stub and safety nozzle during periodical inspection at Tsuruga Power Station Unit 2. Boric acid precipitation was found on the pressurizer relief piping nozzle after the heat insulator was removed. Once the boric acid was removed from the surface of the piping, the area was examined using the SUMP procedure. This is a procedure where a sample of the cracked area is removed and then analyzed. The results of this evaluation revealed a minor crack on the surface of the weld portion of the piping nozzle stub. On the same relief nozzle, cracking was found in an area where boric acid precipitates were not located. This second crack indication was also found in the weld region. Both cracks developed in areas where a large portion of the weld consisted of touch up weld. In another instance, cracking was also located on safety nozzle A.

Reactor Coolant System Nozzles and Other Alloy 600 Cracking

B&W Plants

Arkansas Nuclear Unit 1 (Event Notification Report 36697 + LER 313-2000-003)

On February 15, 2000, a flawed weld was identified on an instrument connection to the reactor coolant system Loop "A" hot leg piping. Once the insulation had been removed, leakage was discovered on five other nozzles. Further investigation using NDE revealed that leakage was occurring through flaws in the partial penetration weld. Both axial and circumferential flaws were found. There was also a subsurface flaw found in a seventh nozzle. Six of the seven level tap nozzles and welds were replaced with Alloy 690. The seventh nozzle weld was repaired using a weld pad buildup and fillet weld.

Westinghouse 4-Loop Plants

Catawba Unit 2 (Event Notification Report 41048)

On September 16, 2004, the steam generator bowl drain for the 2A, 2C, and 2D steam generators were visually inspected. Leakage was found on the 2C and 2D SG bowls. The leakage occurred sometime after the previous refueling outage, because the bowls were clean at that time. Dye penetrant exams were conducted on 2D which identified indications. The root cause of the leakage was determined to be PWSCC. The cracks initiated in the gap between the pipe coupling and the Alloy 600 weld metal buildup. This area is in contact with primary water at a temperature of approximately 617 °F.

Catawba Unit 2 (LER 414-2001-002)

On September 19, 2001, a walkdown of the SG 2B lower head bowl drain indicated boron residue buildup on the ½-inch piping immediately below the SG. The root cause of the SG 2B bowl drain leak was PWSCC of Alloy 600 material. The 2B SG bowl drain was repaired and tested satisfactorily. The remaining three SGs on Unit 2 were visually inspected and liquid penetrant tests were performed. No similar leaks were detected on the other SGs.

Westinghouse 3-Loop Plants

V.C. Summer Unit 1 (Event Notification Report 37423 + LER 395-2000-008 + Final Westinghouse investigation report WCAP-15616, Revision 0, "Metallurgical Investigation of Cracking in the Reactor Vessel Alpha Loop Hot Leg Nozzle-to-Pipe Weld at the V.C. Summer Nuclear Generating Station," January 2001.)

On October 7, 2000, 100 to 200 pounds of boric acid was identified in the "A" hot leg area of the reactor vessel. The potential leak area was identified on the first weld off the reactor vessel at the nozzle-to-pipe connection of the Loop "A" hot leg. . Ultrasonic and eddy current inspection together with visual inspection identified the flaw as axially oriented and about 3 inches in length. The flawed weld was removed and a new weld made of Alloy 52/152 material was utilized. Subsequent cleaning and destructive examination of the salvaged pipe section revealed also a short (about one inch) circumferential crack located in the weld that may have branched from the axial flaw, or may have nucleated as a separate defect.

CE Plants

Palisades Unit 1 (LER 255-1993-009)

On September 16, 1993, plant personnel identified a leak in the power operated relief valve line near the nozzle connection to the pressurizer. The crack initiated in the heat-affected zone of the power-operated relief valve Alloy 600 safe end. NDE and visual inspection revealed a circumferential crack approximately 3 inches in length (about 30 percent of the circumference).

San Onofre Unit 3 (LER 362-1995-001)

On July 27, 1995, radio-chemistry evaluation confirmed that RCS weepage had occurred on two hot leg instrument nozzles. The accessible exterior of the two RCS hot leg nozzles were replaced with new Alloy 690 nozzles. The access to the interior of the RCS hot leg piping prevents welding from the inside of the RCS. Therefore the old nozzles were cut off half way through the RCS hot leg materials and the new nozzles were welded to the exterior of the RCS pipe.

St. Lucie Unit 2 (LER 389-1995-004)

On October 10, 1995, during a routine RCS visual leak check, an apparent boric acid buildup was discovered on the "B" side RCS hot leg instrument nozzle. Further investigation confirmed that pressure boundary leakage had previously occurred. The defective instrument nozzle and other instrument nozzles of the same heat were replaced with Alloy 690.

San Onofre Unit 3 (LER 362-1997-001)

On April 11, 1997, during a routine inspection, four hot leg RCS nozzles were found to have leaks and a fifth was suspected of leaking. It was suspected that the leakage was attributable to cracks through the nozzle in the heat-affected zone of the partial penetration weld on each of the instrument nozzles. The outer half of the Alloy 600 nozzle was replaced with Alloy 690.

San Onofre Unit 3 (LER 362-1997-002)

On July 3, 1997, RCS nozzles were inspected and one hot leg spare RTD thermowell nozzle had an increased amount of white residue. An isotopic analysis determined the residue was boric acid from the RCS. PWSCC was believed to be the root cause of the leaks reported. It was believed that the leakage came from a crack in the heat-affected zone of the partial penetration weld on each of the instrument nozzles. The outer half of the nozzle was replaced with Alloy 690 material using a half-nozzle repair technique.

San Onofre Unit 2 (LER 361-1998-002)

On January 26, 1998, plant personnel visually inspected all RCS nozzles in the hot and cold legs, the pressurizer, and the steam generator channel heads. Seven nozzles were identified for repairs. The leakage from these nozzles was not measurable and the evidence of leakage could not be detected until the RCS insulation was removed. It was believed that the leakage from the nozzles came from cracks in the heat-affected zone of the partial penetration weld of the instrument nozzles. Three of the nozzles were replaced with Alloy 690 using a half-nozzle replacement technique. The other four were repaired using an MNSA technique.

Waterford Unit 3 (Event Notification Report 35407 + LER 382-1999-002)

On February 28, 1999, evidence of boric acid leakage was found on three nozzles. One was on the RCS hot leg number 1 resistance temperature detector (RTD) nozzle, a second was on the RCS hot leg number 1 sampling line, and a third was on the RCS hot leg number 2 differential pressure instrument nozzle. The three hot leg nozzles were repaired using the MNSA technique.

Waterford Unit 3 (Event Notification Report 37442, 37434 + LER 382-2000-011)

On October 19, 2000, during a bare metal inspection, boric acid was found on two of the three MNSA clamps that had been installed on hot leg nozzles during refueling outage 9. These clamps had been installed as temporary repairs until a permanent repair could be made during refueling outage 10. The MNSA clamp leakage could have been caused by the flange not being flat against the pipe. The leak could have also arisen from a brief leakage while the clamps seated. All three MNSA clamps were removed and permanent weld repairs were made on the leaking RCS hot leg nozzles.

St. Lucie Unit 1 (Event Notification Report 37919 + LER 335-2001-003)

On April 14, 2001, leakage was discovered on a pipe-to-nozzle connection on line I-3/4-RC-126. This line was determined to be the "B" RCS hot leg instrument nozzle connection for differential pressure (D/P) transmitter PDT-1121D. The nozzle was replaced using a half-nozzle design.

Arkansas Nuclear One, Unit 2 (Event Notification Report 37199 + LER 368-2000-001)

On July 30, 2003, one RCS hot leg RTD nozzle was found to be leaking. Ultrasonic testing of the RCS hot leg base metal adjacent to the RTD nozzle showed that there were no cracks in the hot leg pipe around the RTD nozzle. The RCS RTD nozzle was repaired with an ASME code-qualified process.

Waterford Unit 3 (Event Notification Report 40277)

During an inspection on October 24, 2003, evidence of leakage was detected on nozzle RC-IPT-0106B, which is a pressure transmitter that taps off of the reactor coolant system hot leg #2. The leakage was located in the annulus area where the nozzle penetrates the head. Nozzle RC-IPT-0106B was corrected with a permanent partial nozzle welded repair.

CE Standard Plants

Palo Verde Unit 1 (Event Notification Report 36256 + LER 528-1999-006)

On October 2, 1999, evidence of an RCS pressure boundary leakage was discovered. The leakage was discovered at two Alloy 600 nozzles, one in each of the RCS hot legs. One was at the nozzle upstream of Valve RCV0285 in the line to a steam generator number 2 differential pressure instrument. The other was at the nozzle upstream of Valve RCV-277 in the line to a steam generator number 1 differential pressure instrument. The leakage was discovered in the form of small deposits of boron accumulated around the circumference of the nozzles. Isotopic analysis of the boron accumulation detected only long-lived radionuclides, indicating that it has taken more than 3 years for the reactor coolant to migrate through the nozzle weld and wall thickness. The repair included cutting of the old nozzle and welding on a new Alloy 690 nozzle to the outside diameter of the hot leg pipe.

Palo Verde Unit 1 (Event Notification Report 37878 + LER 528-2001-001)

On March 31, 2001, during visual inspection of the RCS piping, boric acid residue was discovered on the Alloy 600 RCS hot leg thermowell 1JRCETW0121HB. The visual indications were characterized as white streaks fanning out from the hot leg and continuing up the taper of the thermowell with some buildup on the top of the tapered portion. The repair consisted of cutting off the old Alloy 600 nozzle and welding an Alloy 690 plug.

Palo Verde Unit 3 (Event Notification Report 38332)

On September 20, 2001, evidence of RCS leakage was discovered. The leakage was discovered in an RCS hot leg temperature nozzle. The RCS hot leg nozzle was located in the RTD nozzle for an inservice temperature detector (Loop #1, equipment ID: 3JRCETW112HD). The leakage was identified by the discovery of boron deposits accumulated around the circumference of the hot leg nozzle.

Palo Verde Unit 3 (Event Notification Report 39714 + LER 530-2003-002)

On March 29, 2003, engineering personnel performing preplanned visual examinations of RCS piping discovered boric acid on the hot leg instrument nozzle. There was boric acid precipitation found around the instrument nozzle that penetrates the Loop 1 hot leg. The leaking heater sleeve was replaced with an Alloy 690 nozzle.

Boiling-Water Reactor Plants — All Components

GE Type 4 Plants

Vermont Yankee (LER 331-1999-006)

On April 26, 1986, ultrasonic examination of the N5A and N5B core spray safe-end-to-nozzle welds indicated intergranular stress-corrosion cracking (IGSCC). The IGSCC cracks were located in the nozzle weld butter material, Alloy 182. The indications were predominately axial in orientation with seven indications in each weld. The repair technique utilized a weld overlay procedure.

Hope Creek 1 (LER 354-1997-023)

On September 19, 1997, a leak was discovered on core spray nozzle safe-end weld #N5BSE associated with the "A" core spray subsystem. The N5BSE weld was nondestructively tested in the previous refueling outage. This NDE test had been improperly evaluated and the crack had been unrecorded. The cause of the through-wall leakage has been attributed to IGSCC in the Alloy 182 weld metal. A weld overlay technique was used to repair the leaking core spray nozzle safe-end weld.

Duane Arnold 1 (LER 331-1999-006)

On November 5, 1999, two indications of IGSCC were identified in weld RRB-F002. One indication was approximately 44 percent through-wall and the other was approximately 65 percent through-wall. The inspection was expanded and a 65-percent through-wall crack was found in weld RRD-F002. These two F002 welds were repaired by completing weld overlays using Alloy 52. The cause of the cracking was IGSCC in the Alloy 182 weld metal. A reference LER which may have had similar cracking is 331-1985-010.

Susquehanna 1 (Event Notification Report 40605)

On March 23, 2004, during a routine inspection an indication was discovered on the N1B penetration. This reactor vessel penetration is associated with the reactor recirculation B loop. The crack had been detected during previous outages, however, it was not designated as a crack. The crack was circumferential and was approximately 50 percent through-wall. The length of the crack was roughly 2.2 inches (approximately 7 percent of the diameter). The plant used a weld overlay technique to repair the flaw.

GE Type 3 Plants

Pilgrim 1 (LER 293-2003-006)

On October 1, 2003, reactor coolant leakage was detected in a reactor vessel nozzle-to-cap weld. The crack was contained within the Alloy 182 weld metal. After the nozzle was initially welded to the cap, defects were detected and the weld was repaired. The leakage was believed to be attributable to a crack left in the weld materials during the previous repair procedure. The repair procedure utilized a weld overlay technique with Alloy 52.

www.ingramcontent.com/pod-product-compliance
Lightning Source LLC
Chambersburg PA
CBHW081600170526
45166CB00009B/2764